Workforce Planning for Successful Organization Change

The mission of the Awwa Research Foundation (AwwaRF) is to advance the science of water to improve the quality of life. Funded primarily through annual subscription payments from over 1,000 utilities, consulting firms, and manufacturers in North America and abroad, AwwaRF sponsors research on all aspects of drinking water, including supply and resources, treatment, monitoring and analysis, distribution, management, and health effects.

From its headquarters in Denver, Colorado, the AwwaRF staff directs and supports the efforts of over 700 volunteers, who are the heart of the research program. These volunteers, serving on various boards and committees, use their expertise to select and monitor research studies to benefit the entire drinking water community.

Research findings are disseminated through a number of technology transfer activities, including research reports, conferences, videotape summaries, and periodicals.

Workforce Planning for Successful Organization Change

Prepared by:
PA Consulting Group, Inc. (Formerly Hagler Bailly Services, Inc.)
1750 Pennsylvania Ave, NW, Suite 1000, Washington, DC 20006-4506
and

Demarche Consulting Group, Inc (Formerly Rose Enterprises, Inc.)
and

EMA Services, Inc.

Sponsored by:
Awwa Research Foundation
6666 West Quincy Avenue, Denver, CO 80235-3098

Published by:

Awwa Research Foundation
Advancing the Science of Water®

IWA Publishing

DISCLAIMER

This study was funded by the Awwa Research Foundation (AwwaRF). AwwaRF assumes no responsibility for the content of the research study reported in this publication or for the opinions or statements of fact expressed in the report. The mention of trade names for commercial products does not represent or imply the approval or endorsement of AwwaRF. This report is presented solely for informational purposes.

Published by IWA Publishing, Alliance House, 12 Caxton Street, London SW1H 0QS, UK

Telephone: +44 (0) 20 7654 5500; Fax: +44 (0) 20 7654 5555; Email: publications@iwap.co.uk;
Web: **www.iwapublishing.com**

AwwaRF report number 90965F. Originally published by AwwaRF for its subscribers in 2003.
IWA Publishing version published 2004.

ISBN 1 84339 871 0

CONTENTS

LIST OF TABLES vii

LIST OF FIGURES ix

FOREWORD xi

ACKNOWLEDGMENTS xiii

EXECUTIVE SUMMARY xv

CHAPTER 1: INTRODUCTION AND BACKGROUND 1
Economic and Demographic Factors are Driving Change 1

CHAPTER 2: WORKFORCE PRACTICES AND THE WATER PROFESSION 5
Survey of Water Utility Workforce Practices 5

CHAPTER 3: CHANGE AS A PROCESS 17
Creating Meaningful Work Translates Into Increased Productivity 17
Labor/Management Cooperation is Key to Success in a Unionized Environment 19
A Four-Step Process Provides a Context for Change 21

CHAPTER 4: PRACTICES FOR SUCCESSFUL ORGANIZATION CHANGE 23
Establish a Sense of Urgency 23
Provide and Build Leadership 24
Create Organization Alignment Around a Common Vision 27
Create Commitment by Continuous Participation 28
Delegate Decision Making and Create Accountability 29
Effectively Communicate Continuously 31
Establish and Track Meaningful Measures 31
Develop a Flexible Workforce 33
Provide Appropriate Compensation and Rewards 34
Use Supporting Human Resources Tools 34
Build and Maintain Effective Relations 35

CHAPTER 5: WATER UTILITY "BEST PRACTICE" CASE STUDIES 37
Alignment of an Organizational Performance System: Seattle Public Utilities 39
Participation-Working in Teams: Louisville Water Company 42
Knowledge Transfer: Portland Water Bureau 43
Work Process Redesign-Water Distribution System Excavation: Ottawa-Carleton Water Division 43
Utility Worker Classification-Water Service Utility Worker Apprentice Program: Portland Water Bureau 44
Compensation-Performance Recognition: Louisville Water Company 45
Union Relations: Louisville Water Company 46

APPENDIX A: DESCRIPTION OF WORKFORCE ELEMENTS 49

APPENDIX B: SUCCESSFUL WORK PRACTICE DATA COLLECTION FORMS 55

APPENDIX C: SURVEY RESULTS - BAR CHARTS 73

REFERENCES 137

ABBREVIATIONS AND ACRONYMS 139

TABLES

4.1 Management vs. leadership 26

5.1 Example practices from strategies for organization change............................. 38

FIGURES

2.1 Category: respondent profile QD – job function in the utility 6

2.2 Survey respondent characteristics 6

2.3 Category: alignment 8

2.4 Category: urgency 9

2.5 Category: workforce flexibility 12

2.6 Category: compensation/rewards 14

3.1 Four-step process for organization change 21

4.1 Continuum of manager – non-manager behavior 30

FOREWORD

The Awwa Research Foundation is a nonprofit corporation that is dedicated to the implementation of a research effort to help utilities respond to regulatory requirements and traditional high-priority concerns of the industry. The research agenda is developed through a process of consultation with subscribers and drinking water working professionals. Under the umbrella of a Strategic Research Plan, the Research Advisory Council prioritizes the suggested projects based upon current and future needs, applicability, and past work; the recommendations are forwarded to the Board of Trustees for final selection. The foundation also sponsors research projects through the unsolicited proposal process; the Collaborative Research, Research Applications, and Tailored Collaboration programs; and various joint research efforts with organizations such as the U.S. Environmental Protection Agency, the U.S. Bureau of Reclamation, and the Association of California Water Agencies.

This publication is a result of one of these sponsored studies, and it is hoped that its findings will be applied in communities throughout the world. The following report serves not only as a means of communicating the results of the water industry's centralized research program, but also as a tool to enlist the further support of the nonmember utilities and individuals.

Projects are managed closely from their inception to the final report by the foundation's staff and large cadre of volunteers who willingly contribute their time and expertise. The foundation serves a planning and management function and awards contracts to other institutions such as water utilities, universities, and engineering firms. The funding for this research effort comes primarily from the Subscription Program, through which water utilities subscribe to the research program and make an annual payment proportionate to the volume of water they deliver and consultants and manufacturers subscribe based on their annual billings. The program offers a cost-effective and fair method for funding research in the public interest.

A broad spectrum of water supply issues is addressed by the foundation's research agenda: resources, treatment and operations, distribution and storage, water quality and analysis, toxicology, economics, and management. The ultimate purpose of the coordinated effort is to assist water suppliers to provide the highest possible quality of water economically and reliably. The true benefits are realized when the results are implemented at the utility level. The foundation's trustees are pleased to offer this publication as a contribution toward that end.

Edmund G. Archuleta, P.E.
Chair, Board of Trustees
Awwa Research Foundation

James F. Manwaring, P.E.
Executive Director
Awwa Research Foundation

ACKNOWLEDGMENTS

The authors of this report express their appreciation to the many utilities and government entities and their staff, for their in-kind cooperation with and participation in this project. In particular, we wish to thank Mike Rosenberger and Susan Bailey, Portland Water Bureau; Steve Duncan and Bill Pecsi, Metropolitan Water District of Southern California; Annette Pope and Charlotte Harrison, Cleveland Division of Water; Andre Proulx and Kevin Kidney, Regional Municipality of Ottawa-Carleton; Julie Horton and Mary Wideman-Williams, Seattle Public Utilities; Sharma Young and Julie Frazier, Butler County Department of Environmental Services; Mike Sturgeon and Beverly Irons, Louisville Water Company; Evelyn Ferguson, Charleston Commissioners of Public Works; Lorin Fields, Philadelphia Water Department; Rick Hagar, Suburban Water Company; Jerry Hayes, Phoenix Water Department; Beverly Ingram, Detroit Water and Sewerage Department; Fred Loweree, El Paso Water Utilities; Maxine Mitch, St. Louis County Water Company; Angie Munch, Winnipeg Water and Waste Department; Dave Muto, East Bay Municipal Utility District; David Pitcher, Central Utah Water Conservancy District; GwenDolyn Ruff, Columbus Water Works; Ken Thompson, Irvine Ranch Water District; Diane Patrick, Connecticut Water Company; Jim Coe, Toronto Works and Emergency Services; and Luci Rowe, Orange County Utilities; for participating in a benchmarking survey and representing their utilities in a benchmarking workshop during this research effort, for their timely submittals of research materials and documents, and for their untiring efforts and patience while our researchers tabulated and analyzed the benchmarking data.

We are also grateful to Susan Bailey, Portland Bureau of Waterworks; Joanne Peterson, Dove Albert, and Cyndy Holtz, Seattle Public Utilities; Andre Proulx, Regional Municipality of Ottawa-Carleton; and Beverly Iron and Mike Sturgeon, Louisville Water Company; for permitting their utilities to serve as change process examples for this manual and for participating in the necessary detailed interviews, following the benchmarking workshop.

The authors also wish to thank Emerson Lomaquahu and Robert Allen, the AwwaRF project officers who initiated and finalized our workforce planning and development study, provided guidance and especially Emerson, who was an interactive participant in the benchmarking workshop, and gave insightful comments throughout this project. In addition the advice and help provided by the Project Advisory Committee (PAC) and their interactive participation in the benchmarking workshop were sincerely appreciated. The PAC consisted of: Sharma Young, Butler County Department of Environmental Services; Beverly Ingram, Detroit Water and Sewerage Department; and David Pitcher, Central Utah Water Conservancy District.

The principal authors and editors of this report were Rick Albani, Elisa Speranza and Glenn Nestel of PA Consulting Group, Inc. and Terry Brueck and Gloria Cannon of EMA Services, Inc. They are indebted to other key contributing authors and researchers, especially Alan Ispass, John Cromwell and Gregg Edeson of PA Consulting Group; Michelle Durgen and Mary Hoeve of EMA Services; and Dr. Linda Paralez of Demarche Consulting Group, Inc. The authors bear sole responsibility for the contents of this report.

EXECUTIVE SUMMARY

Workforce Planning for Successful Organization Change introduces a four-step process to manage changes in workforce practices, while meeting the ever-changing needs of water utilities. Utility workforce practices are usually based on executive management's policies and decisions and are provided to "internal" customers within the organization. Workforce practices are generally administered, monitored and modified through the utility's Human Resources (HR) Department.

This manual is not a traditional AWWARF report in that it is not a HR program manual or a "cookbook" compendium of "how to" methodologies such as "how to do a compensation study." It is also not a HR policy manual or a training program for change management. While benchmarking was used to establish current water utility HR practices and needs, the manual is not a survey tool for surveying HR programs and practices. Instead, the outcome of this research is a HR guidance manual designed to lead/align/support successful organization change and the development of a continuous improvement organization culture.

Early on in the manual we describe the leadership requirements for change management. The manual is formatted around a change management process roadmap and describes method, tools and practices utilities can use to align their management and staff around a strategy to achieve long-term goals, competitive performance and establish a continuous improvement culture.

The manual is designed to provide change management guidance to utility directors, general managers, presidents, executive management's staff and human resources directors/staff. It will describe what, why, when and where various tools and approaches are used to support successful change and the criticality of grassroots support from employees. For example, it will describe why competitive compensation and benefits programs and incentives such as gain-sharing and skill-based pay are critical to successful organization change.

The manual will describe how to develop successful leadership and HR strategies to deal with various types of change drivers, including privatization, consolidation, regionalization and organizational restructuring. This manual can meet the needs of all sizes of utilities (large to small), union and non-union labor environments, and various types of governance structures, from publicly-owned departments of local government to private, investor-owned utilities.

TEAMWORK AND INSTITUTIONAL STRUCTURE

Meeting the challenges water utilities face will require teamwork, loyalty and inspired leadership. To accomplish our goals, we first found out what motivates the utilities' workforce and get them to buy into a vision of excellence. For the 22 water utilities that participated in this research, we solicited their opinions through a detailed survey and then invited them to a three-day workshop to discuss the tabulated results of the survey. In keeping with the context of change management, the findings were organized around 11 relevant categories. Four of the 22 utilities were selected by their peers to serve as case study examples of best practices within the water industry. These case studies are written up in detail in Chapter 5 of the manual.

KEYS TO SUCCESS

Throughout the Handbook, we discuss in detail, these 11 categories of workforce practices for successful organization change:

- Establish and maintain a sense of urgency;
- Provide and build leadership;
- Create and maintain organization alignment;
- Create commitment by continuous participation;
- Delegate decision-making and create accountability;
- Effectively communicate continuously (internal and external);
- Establish and track meaningful measures;
- Develop workforce flexibility;
- Provide appropriate compensation and rewards;
- Use supporting tools and techniques; and
- Build and maintain effective relations (with governance, labor, and regulators).

A FOUR-STAGE PROCESS FOR CHANGE MANAGEMENT

After the survey results were discussed at the workshop, then the four-stage process for change management was introduced. This four-stage program leads both the providers and recipients of HR business services through a self-assessment leading to answers to the following:

Stage I: Assess HR Business Services & Processes

What business services are key to the productivity of the workforce and how are they currently performing?

Stage II: Align, Set the Stage, and Compare to Best-in-Class.

What is our utility's vision and strategy for enterprise improvement and how do HR business services fit into this strategy? How do we provide these services today and how does our performance compare to the best-in-class provider?

Stage III: Design the Improvements.

What can we adapt from the best-in-class provider to improve the effectiveness and efficiency of our services?

Step IV: Implement and Continuously Improve.

What is our plan to implement our improved service and how can we measure this improvement?

The four major steps are graphically presented in Figure 3.1 in the manual. Each stage consists of several steps. Based on the size of the utility, degree of existing competitiveness, and willingness and ability of the organization to change, completion of the process can be expected within three-to-five years.

Chapter 4 of the manual describes successful organization change theory and practice in a number of different workforce categories. Each category of practice is summarized, by identifying leading practitioners and relevant publications, and including a summary of the successful practices. While this breakdown into practice categories is helpful to elaborate the various individual practices, for organization change to be successful, a holistic approach must be employed. Therefore, successful organization change will include many if not all of the categories of successful practices. In other words, employing a specific workforce practice by itself may not be successful, even though the same practice may be successful when employed in concert with other workforce practices.

Benefits of Developing a Continuous Improvement Model

Improved HR business services generate two kinds of returns. The first is higher quality and/or lower cost service delivery to internal customers. Second, since the utility workforce supports or enables core services to be delivered efficiently, as utility personnel improve their skills, so does delivery of core services to external customers (utility ratepayers).

So called "open book" management has become a standard in the power business. Global power giant AES Corporation provides a good case study. The company trains, empowers and incentivizes every employee to understand and pursue the company's business goals. People are trained in job-related skills as well as business basics and employees are encouraged to work past their operational goals towards business and financial incentives. Employees are involved in developing strategic goals and business plans, and help determine and track critical indicators. This manual demonstrates how the water industry can apply the same principles as the power industry to improve their organizational culture as well.

Improvement in utility HR business services typically is only one component of the strategy to improve the overall utility's competitiveness. This research demonstrates that employees are the key to a successful transition. Employee training and cross-training programs will be crucial to prepare employees for new technology/automation and for new multi-skilled job classifications that integrate operations and maintenance. Interpersonal and team interaction skills will be important as the supervisory span of control is increased from the current 1:4 average to 1:20 or higher and new management methods such as self-directed work teams are implemented. As employees take on new roles and responsibilities, participating in competitiveness-driven initiatives such as managed competition and regional expansion, rank and file workers will need to develop new management and business skills.

Providing employee feedback is especially critical. Other factors cited in this manual which play into worker productivity include: balancing work and family, opportunities for personal and professional growth, the organization's attitude towards its employees, diversity, relationship with the community at-large and environmental awareness, and even spirituality and tolerance in the workplace.

A Final Word

What's at stake here is public sector leadership in providing a key public service. The culture of a regulated monopoly must change to a culture of an efficient business. This means that everyone must contribute to the success of this business.

Workforce business process improvements cannot be achieved with only top-down management directives. They need to be developed and implemented by staff at the grassroots level of the organization and are best designed and executed by teams. In an independent authority where utility employees provide both core and business support services, leadership comes from utility management and the improve process is relatively straightforward. Unified leadership and commitment to change simplifies team formation and assures organization-wide support of long-term improvement goals. Where the utility is a municipal agency that receives some HR business services from other municipal agencies, leadership must come from both the utility and the business service provider and teams must engage people from both organizations.

Changing the organizational culture is hard and the risks are great. A utility must commit to a systematic approach and a sustained continuous improvement effort. Failure to do this makes the public utility less competitive and opens the door to private alternatives. The payoff for success, on the other hand, means the best possible deal for the customers … a worthwhile goal.

CHAPTER 1
INTRODUCTION AND BACKGROUND

After decades of relative stability, North American water utilities have entered a period of change that is profoundly reshaping the profession. Although their fundamental public health mission—providing a safe, reliable water supply to their customers—is still paramount, water utilities are being driven by regulations, increased security risks, customer demands, and competitive forces toward new performance benchmarks. Many utilities are searching for new organizational deployment and employee skills to adapt their operations and build a bridge to the future.

ECONOMIC AND DEMOGRAPHIC FACTORS ARE DRIVING CHANGE

There are several important factors driving change in the North American water profession on the threshold of the new millenium, including:

- a growing number of publicly-owned utilities considering alternative project delivery mechanisms such as design-build, and/or privatization of major operations and maintenance activities, creating a contestable market and competitive pressure,
- horizontal and vertical integration of the water business, including the acquisition of dominant North American chemical and equipment suppliers by large international private conglomerates,
- regionalization pressures ranging from shared services to consolidating multiple jurisdictions into a single regional authority or commission,
- consolidation and acquisitions of small public and private water companies by larger investor-owned utilities ,
- customer-driven pressures to stabilize rates and fees and improve quality, effectiveness, productivity and accountability of utility operations,
- increasing need for more skilled staff to manage investments in new technology and automated systems,
- aging infrastructure requiring a more comprehensive approach to asset management and financing replacement and rehabilitation, and
- ever-increasing environmental and health regulations requiring investment in technology and creativity in achieving compliance, often with dwindling resources.

Of course, the drivers of change in the water business are occurring within the context of sweeping societal change. According to noted futurists Dr. Gregg Edwards and David Pearce Snyder, institutional trends today and in the near-term future show:

> *"Large private and public sector institutions (will) adopt new social technology, transforming themselves through delayering, outsourcing and employee upskilling from pyramidal, hierarchical, vertically-integrated authoritarian bureaucracies to flat, collaborative networks of authoritative competency, in which essentially all operational decision-making will be*

transferred from managers to 'info-mated' rank and file workers."[1]

By the year 2010, it is predicted that homes, schools, businesses and public agencies will be linked to thousands of information sources, allowing all sorts of decision-critical information to flow into daily life. According to Edwards and Snyder, "The resulting ease of communications between individuals and institutions will not only revolutionize the relationship between employers and employees, but the relationships between producers and consumers, doctors and patients, students and teachers, plus affinity groups and their members, and voters and politicians as well."[2]

The percentage of Americans providing informal care to a relative or friend over 50 is predicted to more than double, in just the next five years. Child care responsibilities are also a factor in worker productivity and well-being. The so-called "sandwich generation" of baby boomers will be most affected, caring for children and aging parents at the same time[3]. As technology enables new ways of doing business, more workers will demand alternative work arrangements such as telecommuting, flex-time, job sharing and part-time work. Water utilities, competing with other sectors for skilled workers, will need to pay attention to these trends in order to remain effective.

> The U.S. population is expected to increase by 50% by the year 2050, and non-white groups will make up nearly half the population by that time. According to the U.S. Department of Labor:
>
> - immigration will account for almost two-thirds of the nation's population growth
> - the population of older Americans is expected to more than double
> - one-quarter of all Americans will be of Hispanic origin
> - almost one in ten Americans will be of Asian or Pacific Islander descent
> - more women and people with disabilities will be on the job.[1]

To remain competitive, water utility workforce practices must respond.

Water utilities are capital intensive by their nature, steeped in a tradition of technology and engineering approaches. Some utilities have therefore approached competitiveness goals through investments in new technology, automated process controls, and information management systems. Many of these utilities become frustrated when these investments alone do not produce the desired productivity improvements. Employees often do not have the necessary skills to operate and maintain the new technology, union contracts can preclude improvements in labor productivity, and current organizational and compensation structures can be impediments to change. If employees do not understand or buy into the competitiveness mission, they react defensively, rather than being committed to and enthusiastic about the new direction.

Employees are the foundation for excellence in any organization, and this is especially true in service organizations like water supply utilities. As water utilities embark on programs to improve their productivity, quality of service, and overall competitiveness, workforce issues are critical to the success of these initiatives. A broad range of workforce planning and development

---

[1] Edwards, Gregg and Snyder, David Pearce, "Roller Coaster 2000: Forces Re-Shaping Daily Life and Work in America," The Snyder Family Enterprise, 2000, p. 2.
[2] *Ibid.*, p. 5.
[3] *Ibid.*, p. ES7.

issues is emerging, from new compensation/reward structures to planning for diversity to revised work management methods to new reporting structures.

Employee training and cross-training programs will be crucial to prepare employees for new technology/automation and for new multi-skilled job classifications that integrate operations and maintenance. Interpersonal and team interaction skills will be important as the supervisory span of control is increased from the current 1:4 average to 1:20 or higher and new management methods such as self-directed work teams are implemented. As employees take on new roles and responsibilities, participating in competitiveness-driven initiatives such as managed competition and regional expansion, rank and file workers will need to develop new management and business skills.

Workforce demographics in water utilities will also drive change. With 50% of the workforce reaching retirement age in the next 10 years, succession planning will be critical. Effective recruitment strategies and methods to screen the competency of job candidates are essential if water utilities are to successfully compete for skilled workers. Employees displaced by downsizing will need retraining and effective redeployment strategies to prepare them for new jobs. Many utility organizations may choose to implement retirement incentive packages if a large percentage of their workforce is nearing retirement age.

Labor relations issues, including negotiating changes in union contracts to deal with difficult issues such as changed work rules, consolidated job classifications, and displaced workers, will require a high level of creativity. Collective bargaining may also break down and create contentious circumstances. These situations require resolution methods such as third party arbitration to avoid labor actions or extended litigation.

The good news is that these workforce planning and development issues have been successfully managed before by best practice water utilities and by many other manufacturing and service sectors of the economy. Many North American-based manufacturing and technology sectors, challenged by foreign competition, have successfully re-engineered their work methods, management structures, and supply chain relationships. In the course of this transition, they have effectively managed the associated workforce issues. Most recently, telecommunications, electric and gas utilities have been forced to manage these workforce transition and development requirements as they prepare to operate in a deregulated, competitive marketplace. Unions and management have found new ways to resolve disputes and work cooperatively to maintain a healthy, competitive enterprise.

Research approach draws on best practices to provide guidance.

There is a wealth of information already available on workforce planning and development methods that can be adapted for application to water utilities. The research approach used to develop this manual included a non-conventional survey and workshop to gather information and feedback needed to make these adaptations. This manual was then developed around and linked with the change process.

This manual is built around three fundamental principles, which formed the context for this research project:

1. The water profession is in transition.
2. Change management is a process and must be endorsed by executive management within the utility if it is to succeed.
3. Employees are the key to a successful transition.

The manual is *not* a human resources program manual or a "cookbook" compendium of "how to" methodologies such as "how to do a compensation study." It is *not* a human resources policy manual or a training program for change management. While benchmarking was used to establish current water utility human resources practices and needs, the manual is *not* a survey tool for human resources programs and practices.

Instead, the manual describes how to develop successful leadership and human resources strategies to deal with various types of change drivers, including privatization, consolidation, regionalization and organizational restructuring. The manual is designed to meet the needs of all sizes of utilities (large to small), union and non-union labor environments, and various types of governance structures, from publicly-owned departments of local governments to private, investor-owned utilities. It describes leadership requirements for change management, and is formatted around a change management process roadmap, describing methods, tools and practices utilities can use to align their management and staff around a strategy to establish a continuous improvement culture, achieve long-term goals, practice competitive performance.

The rest of this manual is organized into the following sections:

- Chapter 2 describes the methodology and presents the results of a survey of workforce practices conducted with the 22 participating utilities.
- Chapter 3 provides background on trends in workplace redesign and labor/management cooperation. This chapter also summarizes a four-step process for organizational change.
- Chapter 4 draws together lessons from many of the foremost authorities on organization change, and describes successful practices according to a number of relevant workforce categories.
- Chapter 5 describes case studies of best workforce practices drawn from some of the water utilities participating in this research project.
- Appendix A contains descriptions of workforce elements to which the manual refers.
- Appendix B contains the data collection forms from our four case study utilities.
- Appendix C contains the bar charts generated by the 22-utility survey respondents, which formed the basis for much of the research.

CHAPTER 2
WORKFORCE PRACTICES AND THE WATER PROFESSION

As part of this research project, a Survey of Workforce Practices was administered to the 22 participating utilities. Survey facilitators at each utility administered surveys to their colleagues, collected the results along with utility profile data, and attended a three-day workshop with other participating utilities. Some of the participating utilities were well along the change process, while others were just starting.

The survey was non-conventional in that it was given to a group of water utility employees in each of 22 utilities in a scheduled session. Each employee expressed his or her own opinion about each of the practices described in the survey. Their perspectives depended on their personal experiences and other factors such as their job position, years of tenure, previously held positions and their personal involvement in any change process. The workshop then brought all 22 utilities together to discuss the tabulated results, and benchmark each individual utility against the statistical mean and median scores of all utilities combined.

Utility Participants

Butler County Dept. of Environmental Services
Central Utah Water Conservancy District
Charleston Commissioners of Public Works
Cleveland Division of Water
Columbus (GA) Water Works
Connecticut Water Company
Detroit Water & Sewerage Department
East Bay Municipal Utility District
El Paso Water Utilities
Irvine Ranch Water District
Louisville Water Company
Metropolitan Water District of Southern California
Orange County Utilities
Ottawa–Carleton, Regional Municipality of
Philadelphia Water Department
Phoenix, City of
Portland (OR) Water Bureau
St. Louis County Water Company
Seattle Public Utilities
Suburban Water Systems
Toronto Works & Emergency Services
Winnipeg Water and Waste Department

SURVEY OF WATER UTILITY WORKFORCE PRACTICES

Step 1: Cross-functional survey group defined.

The designated facilitator selected and assembled a diagonal cross-sectional group of participants representative of several functions within the organization. The groups included representatives of the utility's executive management and a cross-section of middle managers, first-line supervisors and other staff. Representation was encouraged from operations, maintenance, engineering, capital planning and administrative/support functions such as human resources, customer service, information systems, finance, purchasing, laboratory, safety, and public affairs. Some utilities also included representatives of organized labor.

Ideal candidates for the survey group were "thought leaders" in the utility, regardless of whether they were advocates or opponents of change. Facilitators were also encouraged to select employees with varying lengths of service. Individual survey responses were used without attribution to preserve anonymity.

Step 2: Survey group assembled and survey conducted.

Facilitators were encouraged to administer the surveys in a group session, with no advance preview or discussion. This method is most time-efficient and assures that each participant's responses are their own individual opinion, made privately and without other influences. Once completed, surveys were placed in sealed envelopes to assure anonymity and sent directly to the research team. The exercise generally took about one hour.

Step 3: Survey results compiled.

The research team analyzed each utility's results and provided each participating utility with a summary report of its aggregate survey results. The survey respondents were roughly split between administrative and operations and maintenance staff, as shown in Figure 2.1. Other respondent characteristics broke down as shown in Figure 2.2:

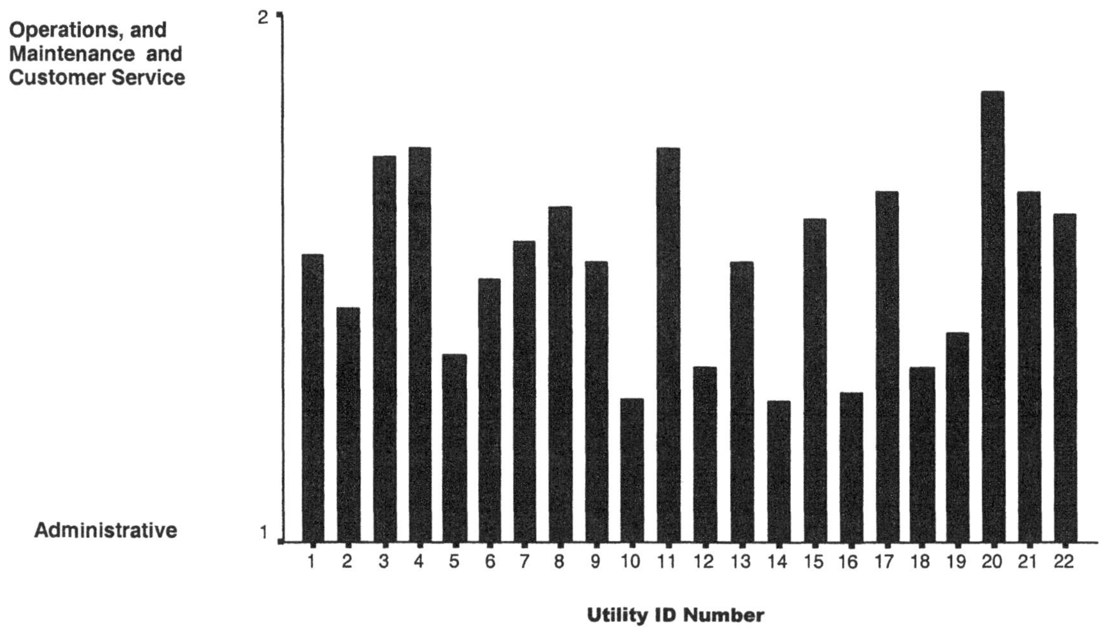

Figure 2.1 Category: respondent profile QD - job function in the utility

Survey respondents characteristics
Non-union:76%, Union : 24%
Senior Managers: 21% Middle Managers: 51% Rank and File: 28%
Administrative: 48% Operations and Maintenance: 52%
> 10 years with the utility: 55% 3-10 years :36% < 3 years: 9%

Figure 2.2 Survey respondents characteristics

Step 4: Survey group assembled to discuss results.

Once the utility's results were returned, the facilitator was encouraged to distribute the report and schedule a second session with the cross-functional group. At this session, which in some cases included additional participants, the survey summary results were discussed and input was received in preparation for the facilitator's presentation at the workshop. Specific planning instructions and a presentation template were included, and facilitators were asked to share: (1) examples of surprises or unexpected findings in the survey results, (2) examples of well-developed practices they believed were important for successful organizational change or workforce development, (3) examples of the most critical areas or barriers to change at their utility and (4) input on how they expected to use the resulting guidance manual.

Step 5: The workshop.

During the workshop, each utility was asked to use the information developed in Step 4 in breakout sessions and summary presentations. Participants were provided with a composite summary of the survey results. Prior to the breakout groups, the research team presented bar graphs of the tabulated results of the combined 22 utilities (included in Appendix E). Each question was discussed with the participants and the analyses of the results were interpreted as well. One of the AWWARF Project Officers and all of the PAC members were also participants in the workshop.

Survey Results

The survey responses were compiled and analyzed by the research team, and the findings are summarized below. In keeping with the context of change management, the findings are organized around 11 relevant categories.

Alignment

Mission, Vision, Values

Most of the 22 utility respondents have established and communicated a mission, vision and values, but viewed themselves as less developed in applying these mission and values to day-to-day utility practices. They also ranked quality management/problem-solving methods as a major factor in achieving their utility's goals.

Long Term Goals and Commitment of Resources

The respondents "somewhat agreed" that their utilities emphasize the creation and communication of long-term goals and apply resources to achieve them. They did however, believe it was very important to link individual goals with organizational goals.

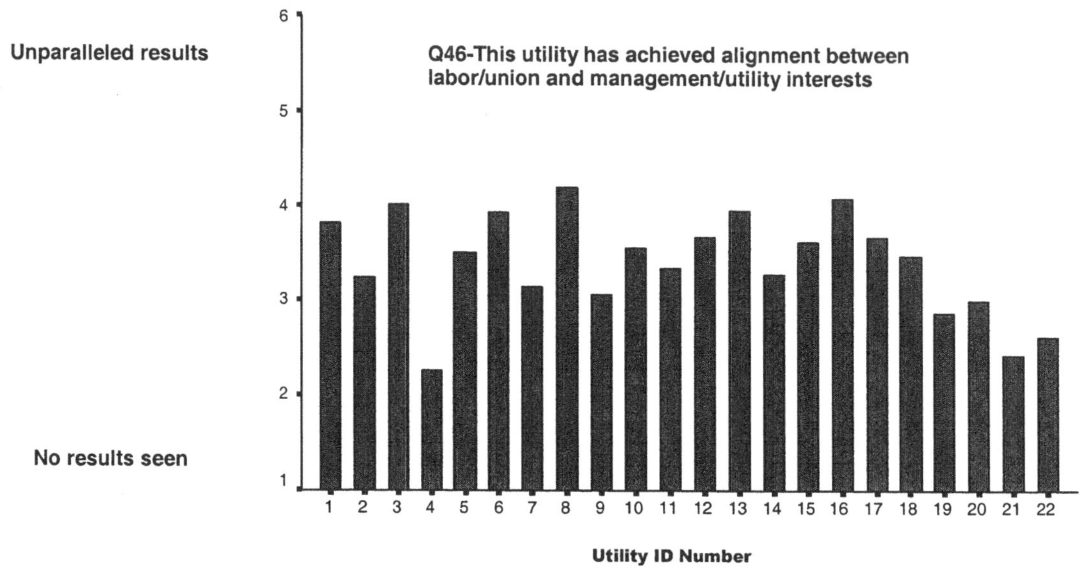

Figure 2.3 Category: Alignment

Customer Satisfaction

The respondents ranked themselves as generally emphasizing individual employee roles in delivering high levels of customer service and linking employee job responsibilities with achieving customer satisfaction.

Management/Labor Alignment

The utilities ranked themselves as below expectation in achieving alignment between the unions and the utility's business interests. Figure 2.3 shows the distribution of responses by utility. The overall mean score is 3.4 in response to the question of achieving alignment between labor / union and management / utility interests, where a rating of 6.0 is equivalent to "unparalleled results" and 1.0 is "no results seen.

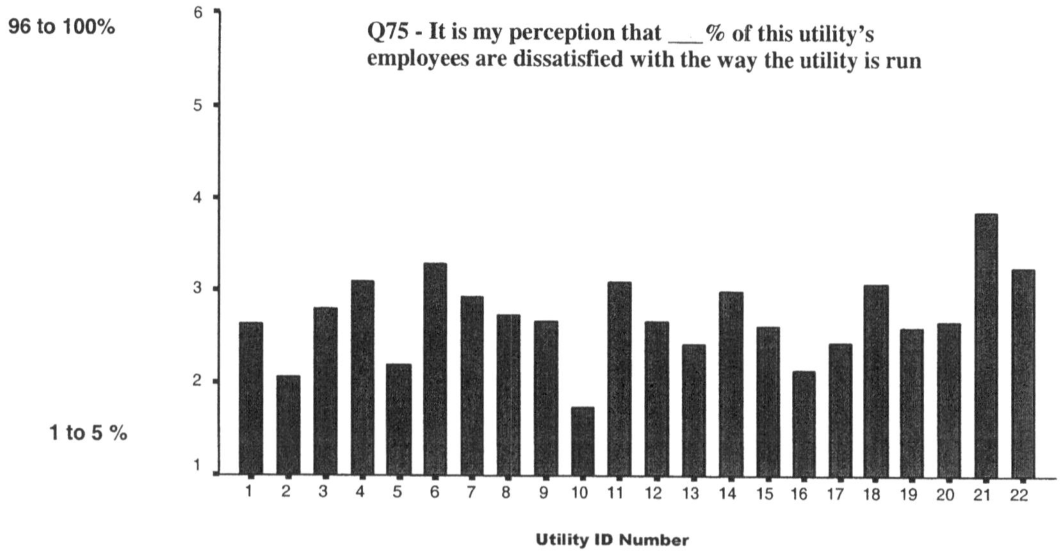

Figure 2.4 Category: Urgency

Urgency

Awareness of Market/Stakeholder Drivers of Change

Respondents indicated there is a high level of awareness and urgency regarding the competitive and stakeholder drivers of change.

Importance of Performance Improvement Plan and Redesigning Work Practices

Respondents ranked a performance improvement plan as very important and indicated a high level of urgency for redesigning work practices to improve performance and customer satisfaction.

Performance Improvement Change Initiatives

Most of the utilities indicated they were already engaged in performance improvement initiatives.

Customer/Employee Dissatisfaction

While the utilities ranked the overall level of customer dissatisfaction as low, respondents indicated a higher level of employee dissatisfaction with the way their utilities are being run. In this case, shown in Figure 2.4, a mean score of 2.7 indicates that this average response on a six-point scale equates to a respondent perception that about 45 to 50 percent of employees are dissatisfied with the way the utility is run.

Leadership

Leadership Commitment and Development

While most of the responding utilities indicated they have highly visible management commitment, they ranked the development of leadership skills as only somewhat emphasized.

Leadership: "Walking the Talk" and Mentoring

Utility senior management is generally committed to "walking the talk," mentoring was ranked as only somewhat used as part of employee's day-to-day job responsibilities.

Impacts on Employees

Utilities were somewhat skeptical regarding senior management's consideration of impacts on employees.

Decision-making

Level of Decision-Making

Utilities indicated they were generally aware of their organization's decision-making process.

Individual Responsibility/Accountability

Utilities indicated that individual employees were "somewhat" willing to make decisions.

Delegation of Authority for Decisions

While the 22 responding utilities considered delegation to the lowest possible level an important success factor, delegation of decision-making was ranked somewhat below expectations.

Participation/Style

Employee Level of Participation

Respondents ranked employee involvement and level of participation in their utility performance improvement efforts as generally good, with roles and responsibilities fairly well defined. Employees generally considered themselves part of the improvement team.

Labor/Management Collaboration and Participation

Labor/management participation in setting up new job classifications, pay scales and performance incentives was ranked somewhat below expectations. This included cooperation in resolving grievances and in dealing with change issues in contract negotiations.

Employee Personal Initiatives

Respondents indicated they generally emphasized and encouraged individual employee initiatives.

Team-based Approach

Use of team-based approaches for problem solving and work process improvement was ranked as only somewhat developed.

Communications

Voice of the Employee: Communications process for employee input and feedback

Utilities indicated they have a "somewhat" developed system of communications supporting the change process, including the process for receiving employee input, suggestions on concerns and providing feedback.

Voice of the Customer: Communications process for customers/stakeholders

Utilities indicated they have a fairly well developed communications process for customers and other stakeholders.

Effectiveness of Communications on Long Term Goals

Communications on long-term goals is somewhat below expectations.

Formal Communications Plan

All 22 utility respondents viewed a formal communications plan as very important, but ranked the current level as only partially implemented.

Communicating/sharing lessons learned

The communications process for capturing and sharing successes and lessons-learned was ranked as fairly well developed.

Worker Flexibility

Confidence in Tool/Training Investment

The respondents gave a high ranking for utility commitments and current investments in tools and training to achieve performance improvement and customer satisfaction goals.

Broader/Multi-skilled Job Classifications

The utilities reported mixed success in transforming their utility to a cross-trained workforce with higher skill levels.

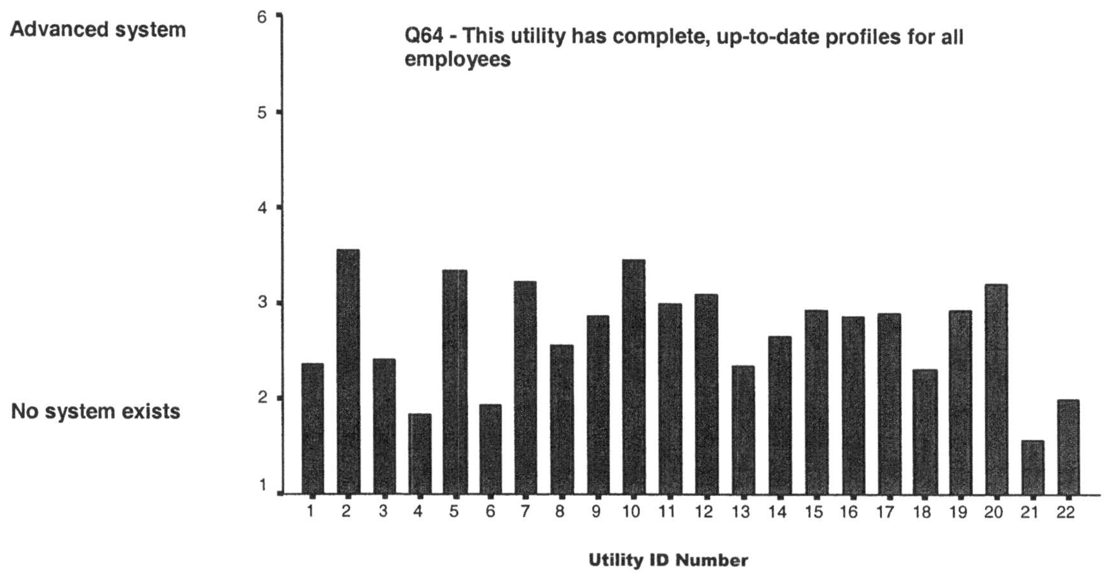

Figure 2.5 Category: workforce flexibility

Multi-skilled Workforce: Level of Implementation

The utilities have achieved some success in migrating from a workforce comprised of narrowly defined job classifications to a multi-skilled workforce with broader, more highly skilled jobs. All indicated some level of multi-skilled workforce deployment.

Multi-skilled Workforce/Cross-training Curriculum and Program

Utilities indicated that the cross-training program has only been somewhat effective.

Employee Skill/Experience Profiles

Respondent utilities indicated they have basic systems in place to maintain up-to-date information on education, training, skills, competencies and certifications. Complete and accurate employee profiles are one of these basic systems in development for most respondents, given the mean score rating of 2.7 shown in Figure 2.5.

Diversity Goals

Diversity goals are only somewhat linked with cross-training programs to expand skills and knowledge of individual employees.

Succession Planning

Succession planning for management and key positions is only somewhat developed among the utility respondents.

Measures

Performance Measurement System for Business Processes

All utilities ranked performance measurement as a critical success factor for performance improvement.

Individual Performance Measures Linked with Contributions to Utility-wide Goals, Effectiveness, Efficiency, Quality and Customer Service

All utilities ranked individual employee performance as very important to overall utility performance. They indicated some degree of correlation between individual and utility-wide performance goal, with some having specific performance measures of individual performance objectives linked with utility-wide goals.

Utility Performance Measures (Effectiveness, Efficiency, Quality, Customer Service)

The utilities indicated that they generally linked measures of business processes and work practices with overall performance and customer satisfaction. Over two-thirds indicated that they used specific measures to track performance and that considerable progress had been made over the last year.

Degree of Impact for Performance Improvement/Change Management over last year

There was a wide variety of responses on the degree of personal employee impact from utility change initiatives over the last year, but the majority ranked these initiatives as having a very significant impact.

Customer Service Performance Measures

Only some of the utilities had specific performance measures of customer service and satisfactions.

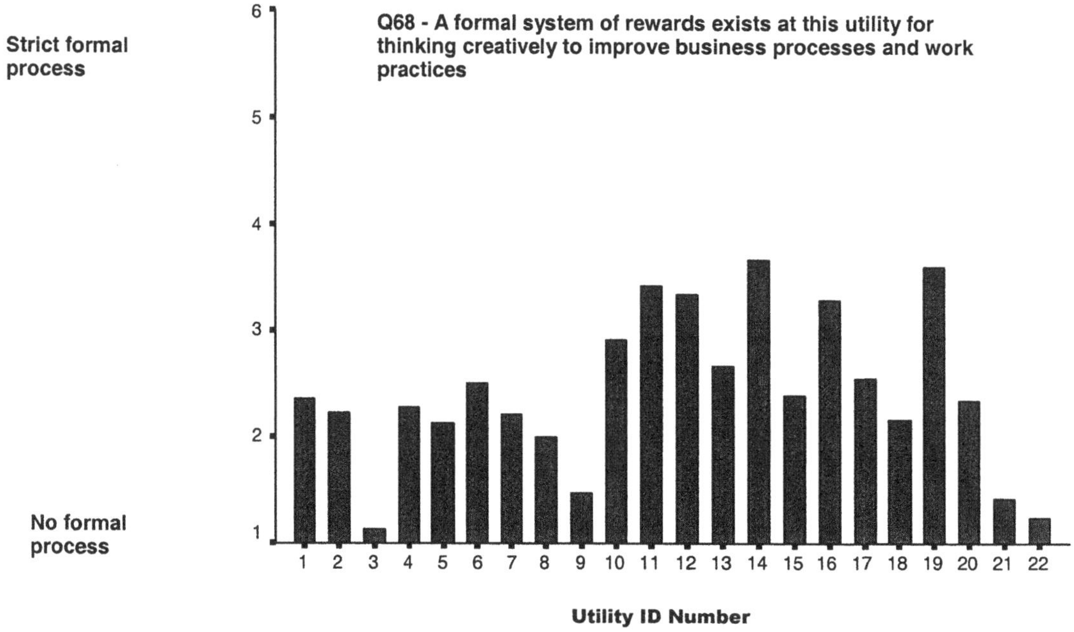

Figure 2.6 Category: Compensation/Rewards

Benchmarking/Comparisons with other Utilies

Considerable investment in early benchmarking and performance comparisons are being done with other utilities. There was less evidence of results and outcomes from this investment.

Annual Management Training

Management receives an average of two-to-five days per year of training.

Annual Employee Training

Most of the respondents indicated an average of two-to-five days per year of individual employee training.

Compensation/Rewards

Formalized Individual Employee Performance Appraisal Process Using Quantitative/Qualitative Criteria

There was a wide variation in the formality of the performance improvement system for individual employees among the respondents. The majority indicated a semi-formal process, as illustrated in Figure 2.6.

Non-Performance-Based Cost of Living (COL) Increases Job Security

There was mixed but fairly strong support for COL-driven increases and job security, without regard to utility performance.

Compensation based on Utility Performance

There was even stronger and more uniform support for individual incentive-based compensation based on overall utility performance.

Compensation based on Team Achievements

There was very strong support for team-based incentive compensation.

Reward/Recognition Programs

While there was generally some level of non-monetary reward/recognition program among the respondents, the programs linked with creative process/performance improvement ideas and contributions were less developed and far more varied.

Policies/Procedures/Tools

Hiring Process/Procedures

The utility respondents showed a mixed response but generally have a formalized process for identifying critical hiring needs. There was also had a mixed response for the actual hiring process for new staff, although this was more developed. Most fell into the category of a three-to-six month cycle time for hiring new staff.

Personnel Grievance/Issues/Conflict Resolution

A formalized, effective process for handling personnel issues also yielded a mixed response.

Employee Health Care/Wellness Programs

Healthcare/wellness programs at the 22 utilities were also mixed.

Information Systems for Human Resources

Most utility respondents were using manual systems.

Change Management Process

Most utilities indicated they had established at least a rudimentary process for managing change issues.

External Forces

Regulatory/Compliance

This category was ranked uniformly by all 22 utility respondents as the most dominant external factor.

Governing Body/City Council

Governing body issues ranked a close second in terms of dominant external factors.

Technology Drivers for Change

There was strong concurrence that technology requirements are creating a compelling case for change and work practice/job redesign.

Factors other than skills, abilities and job performance

Utilities indicated a mix of factors that might be involved (e.g., bumping tied to seniority/labor agreements, civil service, patronage, etc.).

Effective Information Systems for enabling continuous improvement

Although identified as a critical need, current development was ranked low overall.

CHAPTER 3
CHANGE AS A PROCESS

CREATING MEANINGFUL WORK TRANSLATES INTO INCREASED PRODUCTIVITY

Meeting the challenges water utilities face will require teamwork, loyalty and inspired leadership. To accomplish our goals, we must find out what motivates our workforce and get them to buy into a vision of excellence. In a recent survey of 1600 U.S. workers, employees' identified the top three most important aspects of their work. Fifty-two percent wanted to be responsible for their work and the results it produced, 42% wanted acknowledgement for their contributions and 39% wanted their tasks to match their strengths (some wanted more than one of these).[4]
Some key elements for redesigning work include letting employees know:

Seven skills for future workers

- Technology
- Vision
- Measurement
- Organizational
- Persuasive
- Communications
- Learning

- they are a valued part of the whole, by sharing information and helping employees feel responsible for the results,
- their work makes an impact, by directly involving them with process improvements and encouraging contributions and collaborations,
- they are responsible for outcomes, by giving them clear goals and the authority to make things happen,
- their work will be measured, by providing direct and timely feedback so employees can gauge their impact and improve,
- they will be rewarded meaningfully (even though public utilities might have to be more creative with rewards due to resource, legal and contractual restraints), and
- they are encouraged to work as a team, and rewarded accordingly.

The feedback and improvement point is especially critical. As one observer has noted, "working without feedback is like bowling in the dark." Other factors cited in the literature which play into worker productivity include: balancing work and family, opportunities for personal and professional growth, the organization's attitude towards its employees, diversity, relationship with the community at-large and environmental awareness, and even spirituality and tolerance in the workplace[5].

"Open Book" HR Principles

- Hiring for attitude, not aptitude
- 360-degree performance reviews
- Training with innovative tools (games, etc.)
- Variable compensation plans (bonuses pegged to performance

"The top employees of the coming century will be flexible, creative and motivated toward making a positive

[4] Herman, Roger E. and Gioia, Joyce L., "Making work meaningful: secrets of the future-focused corporation", *The Futurist*, December 1998, p. 24.
[5] *Ibid.*

difference in the world. They will seek balance, growth and fulfillment in both their work and home environments. The (organization) of the future must respond to these needs and desires; otherwise, they will find themselves hampered by a lack of qualified people to accomplish the organization's work."[6]

Other utility industries, such as power, gas and telecommunications, have had somewhat of a head-start on water utilities in grappling with changing workforce issues. Successful utility organizations have transformed themselves based on team management skills and visionary leadership for change. Case studies point to the compression of tall hierarchical structures, which have yielded immediate and visible improvements, including increased efficiency, morale and trust between employees and leaders, reductions in cycle time of group processes and decision-making, absenteeism and workers' compensation claims. Organizational goals are clearly defined, and individuals are shown how they fit into the corporate mission.[7]

Transformations at these "best practice" utilities did not take place overnight, and were not generic. Each organization—and in most cases each individual facility—mapped a unique action plan. The unifying theme in each case, however was training and certifying employees to become leaders and, in turn, to train their colleagues.

Faced with suspension by the U.S. Nuclear Regulatory Commission in 1993, Houston Lighting & Power Company's South Texas Project (the second largest nuclear plant in the United States) needed a drastic overhaul. They shifted from a command-and-control hierarchical style of management to a "facilitative leadership" program, which redefined the way employees worked with each other, suppliers and customers. The company's continuous improvement model was incorporated into a customized plan, which trained senior managers and then cascaded the training throughout the organization. By training line managers—who had the most credibility with the workforce—the company accomplished quick change, established a common language and managed to implement a critical cultural shift. The project not only survived the suspension threat, but achieved top performance status in efficiency, operations and cost[8]. The South Texas model has been replicated throughout the United States as other threatened nuclear plants strive to improve safety and performance.

Drivers of change in public workforce management

- growing demand for quality services
- new technology changing the way people live, work and communicate
- financial pressures
- traditional budgeting, personnel and labor relations procedures hampering change
- environmental pressures
- increase in unionization and more active public employee unions supporting workplace innovation
- highly educated public workers wanting a say in workplace decisions
- workers and management seeking cooperative ways of handling issues, moving away from traditional conflict-based systems
- increased interest in contracting out to reduce costs and/or improve service

---

[6] *Ibid.*
[7] Jurow, Doree, "Build Management Teams with the Power to Lead", *Power*, March/April 1998, p. 79.
[8] *Ibid.*

So called “open book” management has become a standard in the power business. Global power giant AES Corporation provides a good case study. The company trains, empowers and incentivizes every employee to understand and pursue the company’s business goals. People are trained in job-related skills as well as business basics and employees are encouraged to work past their operational goals towards business and financial incentives. Employees are involved in developing strategic goals and business plans, and help determine and track critical indicators.

Change management was implemented in small, achievable steps. As one observer noted, “One sure route to failure is to launch a change initiative with a huge amount of fanfare, implement it from the top down and expect behavior to change instantly.”[9]

At AES, “One of the only centrally-controlled aspects ... is our commitment to four major "shared" principles: to act with **integrity**, to be **fair**, to have **fun** and to be **socially responsible**. These principles are goals and aspirations to guide the efforts of the people of AES as we carry out the mission of the Company.”[10] Shared values such as these are woven throughout every aspect of successful organizations and are some even print them on the back of business cards so each employee has a constant reminder of where they are headed.

Water utilities are mostly not facing the threats as the nuclear power industry, nor are they undergoing the type of dramatic restructuring brought about by utility deregulation (at least not yet). There are, however, some valuable lessons to be learned about the impact of change management programs on efficiency and customer service.

LABOR/MANAGEMENT COOPERATION IS KEY TO SUCCESS IN A UNIONIZED ENVIRONMENT

In 1996, the U.S. Labor Secretary Robert Reich issued a report from his Task Force on Excellence in State and Local Government through Labor-Management Cooperation called *Working Together for Public Service*. The 14-member Task Force was drawn from the ranks of labor, management, elected officials, neutrals and academics.

The Task Force identified typical barriers to establishing workplace partnerships between unions and management which will have a familiar ring to water utilities, including: historic mistrust, lack of communications skills, failure to include all affected parties, refusal to try new approaches, fear of job loss, fear of union busting, and managers and/or officials who feel their status is threatened by a team-oriented approach.

Developing, implementing and sustaining workforce partnerships with management require the absolute commitment from all parties, patience and perseverance and almost always changes to the underpinning systems which serve the organization’s core functions. In order to overcome traditional barriers, the Task Force recommends starting small, building trust, making decisions at the local level, ensuring the absolute commitment of leadership, breaking with the past through training and conflict resolution, assuring employment security, respecting the role of the union, remaining flexible, paying attention to jurisdictional issues, and reducing hierarchical decision-making.

While not all of their observations and recommendations are necessary applicable—or desirable—for all water utilities, the Task Force did unearth some interesting and instructive case studies. Their findings were also consistent with many of the best practices our research team drew from both inside and outside the water business, including:

---

[9] Case, John, “HR Learns How to Open the Books,” *HR Magazine*, May, 1998.
[10] AES web site.

- the need to change the way the public workplace is managed and the way the public employee is engaged in the process,
- the need to alter traditional methods of service delivery, personnel and administrative systems, supervision and workplace communication, and approaches to collective bargaining to meet the challenges of the future,
- the imperative of employee participation to achieve continuous improvement goals and reducing conflicts, particularly in collective bargaining situations, and
- the need for new tools to move from confrontation to cooperation among management, union officials and workers.[11]

After examining dozens of examples across the United States, the Task Force concluded that when employees are engaged in service planning and implementation decisions, the results are reflected in better, more cost-effective service, a better quality of work life and improved labor-management relations. Case studies examined during the Labor Department's investigation bear out this conclusion. A few examples:

- The State of Ohio's Civil Service Employees Association (OCSEA/AFSCME Local 11) and the State began implementing a quality improvement program in 1991. A Steering Committee made up of cabinet officials, agency directors, and union officials oversaw the effort, drawing upon the expertise of private sector practitioners, studying quality improvement tools and techniques and developing a mission statement and implementation plan. This major effort was rolled out to all state employees, and successful activities are chronicled in a "results" book. The continuous improvement program is enshrined in Article 21 of the union's contract with the state, and works outside the every day issues of collective bargaining, such as salaries, benefits, contract interpretations, disciplinary actions, and grievances. Instead, it addresses issues of operational efficiency and quality, including work process improvements. The program was hailed by the *Washington Post* in 1997 as "one of the best examples of labor-management partnership in government" and is the subject of a Case Study in Public Policy and Management at the John F. Kennedy School of Government at Harvard University.[12]
- In Los Angeles, the Bureau of Sanitation's partnership with SEIU Local 47 improved cooperation between drivers and managers, increasing truck availability by 20% and reducing overtime by 54%. In the face of a budget crunch in the mid-1990s, the City achieved a 25% reduction in the department's costs, without layoffs.[13]
- Phoenix, Arizona's Water Services Department (WSD) saved millions of dollars in chemical, electricity and labor costs through a joint program with its two unions called "PALM"—Participative Association of Labor and Management. Spurred by interest from private contract operators, along with an extensive planned capital program, WSD implemented the three-phase re-engineering process after scrutinizing its organizational structure, work practices, and information and control technology infrastructure. A key element of the program was communication. A PALM newsletter, toll-free information hotline and monthly information meetings at all sites,

---

[11] U.S. Department of Labor, *Working Together for Public Service*, 1996.

[12] *Ibid.*, AFSCME web site, and Osborne, D., and P. Plastrik. March 9 1997. The Union Solution. *The Washington Post*, W08.

[13] U.S. Department of Labor, *Working Together for Public Service,1996.*

along with information kiosks and special events, were used to help explain the initiative and keep employees and other stakeholders up-to-date. [14]

A FOUR-STEP PROCESS PROVIDES A CONTEXT FOR CHANGE

Achieving the objectives of employee buy-in and labor cooperation should be accomplished within the context of an overall change management process. Based on the experience of the research team and participating utilities, a four-step process for organization change provides a methodology for continuous improvement.

Four major steps of an organization change process are presented in Figure 3.1. Each stage consists of several steps. Based on the size of the utility, degree of existing competitiveness, and willingness and ability of the organization to change, completion of the process can be expected within three-to-five years. When combined with the seven keys, however, experience has shown that the process will deliver immediate changes in staff awareness and will also result in immediate productivity improvements.

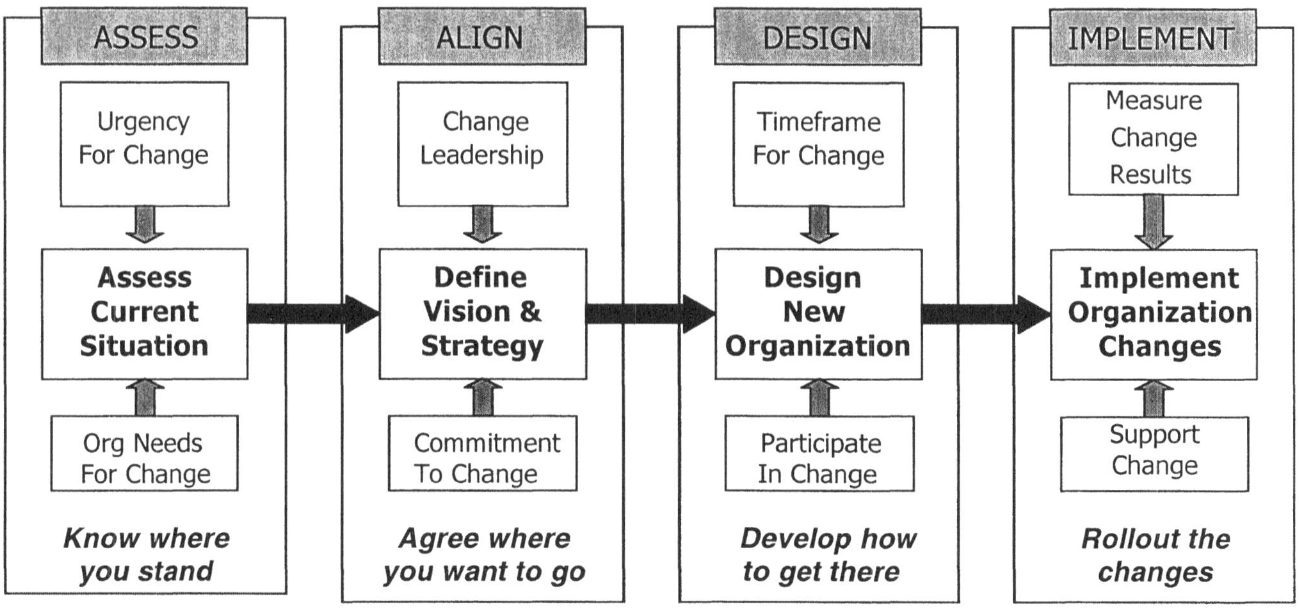

Figure 3.1 Four-step process for organization change

While the process generally occurs sequentially, unique local conditions and the nature of each stage may permit concurrence of certain activities. Generally, stages I and II occur first and in that order to build an appropriate foundation for the remaining stages. Stages III and IV may run somewhat concurrently especially with respect to the implementation of "quick wins." The fifth stage of continuous improvement is an on-going process that, by its nature, may be

[14] *American City & County,* September 1998

concurrent with all other stages. Each stage is previewed below and discussed in more detail in Chapter 4.

ASSESS: Know Where You Stand

This step consists of a range of assessment activities to better understand current conditions relative to change drivers and change urgency including factors of competitiveness, local political environment, customer perception and support, and financial health and stability. This step is crucial for building appropriate awareness of the need for change among staff and other stakeholders.

ALIGN: Agree Where You Want To Go

This step consists of establishing, communicating, and achieving consensus regarding goals as an organization. It includes the crucial step of assembling a "guiding coalition" to lead the organization to its new vision and goals. Vision and mission are identified, confirmed, clarified, or updated. Organizational strategies are determined and prioritized in conjunction with associated performance goals.

DESIGN: Develop How To Get There

Based on the current situation assessed in Step 1 and the broad agreement on goals in Step 2, the organization is positioned to design the major changes required to achieve these goals. Step 3 includes re-assessing change readiness, forming design teams, designing the new organization, developing a change rollout plan, and obtaining governing body support.

IMPLEMENT: Roll out the Changes

This step implements the organization design completed during Step 3. While details of this step will be unique to each organization, success generally results from carefully integrated changes in the structure, grouping, and management of human resources to effectively drive or support the new organization. Examples include training, position descriptions, skill certification, compensation systems, and incentive programs.

Because they are interdependent, a change in an area of practices generally results in changes in organization support. For example, a change in operations philosophy to include certain maintenance work in the routine of plant operations staff, will ultimately require operations staff to be trained sufficiently to complete select maintenance tasks. As a result, position descriptions or compensation systems may be revised and process monitoring and control technology may be modified.

Steps 3 and 4 will likely consist of multiple, parallel transition activities ranging from small pilot projects to larger organization-wide changes. Pilot projects are particularly useful as they permit experimentation to quickly demonstrate success and to improve design details before changing the organization throughout the utility.

CHAPTER 4
PRACTICES FOR SUCCESSFUL ORGANIZATION CHANGE

This Chapter describes successful organization change theory and practice in a number of different workforce categories. Each category of practice is summarized, by identifying leading practitioners and relevant publications, and including a summary of the successful practices. While this breakdown into practice categories is helpful to elaborate the various individual practices, for organization change to be successful, a holistic approach must be employed. Therefore, successful organization change will include many if not all of the categories of successful practices. In other words, employing a specific workforce practice by itself may not be successful, even though the same practice may be successful when employed in concert with other workforce practices.

For instance, in one utility, a proposed change to all job classifications and job composition meets extreme resistance from the workforce. In another utility, this same proposal is immediately found to be acceptable by a strongly positioned, unionized workforce. Why the difference?

In the first case, if the prerequisite understanding of the need for change and subsequent alignment around the new workforce structure has not preceded the job changes, acceptance of the new job classification system will be minimal. In the second case, if the sense of urgency, leadership, alignment, participation, communication and commitment to new job classifications has been understood and acknowledged, acceptance of the changes will be immediate.

The remainder of this Chapter summarizes the practices for successful organization change broken down into eleven categories.

- Establish and Maintain a Sense of Urgency
- Provide and Build Leadership
- Create and Maintain Organization Alignment
- Create Commitment by Continuous Participation
- Delegate Decision-Making and Create Accountability
- Effectively Communicate Continuously (Internal and External)
- Establish and Track Meaningful Measures
- Develop Workforce Flexibility
- Provide Appropriate Compensation and Rewards
- Use Supporting Tools and Techniques
- Build and Maintain Effective Relations (with governance, labor, regulators)

ESTABLISH A SENSE OF URGENCY

In John Kotter's *Leading Change,* establishing a sense of urgency is the first of an eight-stage process toward creating organization change[15]. Likewise, Rosabeth Moss Kanter, et al. established "The Ten Commandments" for executing change, one of which is to create a sense of urgency[16]. This critical step in leading a change effort is easy to advocate when an organization

---

[15] Kotter, John P. *Leading Change*. Boston, MA: Harvard Business School Press. 1996.
[16] Kanter, Rosabeth Moss., B.A. Stein, R.D. Jick. *The Challenge of Organizational Change*. New York: The Free Press. 1992.

is seriously and openly in trouble due to threats such as financial failure or privatization. It is significantly more difficult to change if there is no compelling reason.

Urgency is the extent to which every employee in the organization understands the need for change, and the potential consequences if the status quo is maintained. Creating, supporting, and driving this sense of urgency requires bold and sometimes risky moves by organization leaders. Some factors involved in creating urgency to change include:

- Awareness of external market/industry forces
- Threats of privatization
- Political drivers
- Customer dissatisfaction
- Employee unrest
- Failures (equipment, plant, people)

Utility leadership must be aware that complacent and over-cautious staff including managers, who are used to a culture of having everything under control, can consciously or unconsciously sabotage major transformation efforts. Many efforts to change are "over managed and under led,"[17]. This reference to a major difference between managers and leaders is important to note. In essence, leaders set the pace and create direction, while managers keep things running. Responsibility for maintaining and communicating the sense of urgency to change falls on organization leaders.

Sharing information for future opportunities as well as disclosing unfavorable data or trends helps people to see things with a realistic perspective. Keeping employees "in the loop" by honestly sharing good news *and* bad news creates trust and enforces belief in the sense of urgency.

PROVIDE AND BUILD LEADERSHIP

Building leadership demonstrates the commitment of the utility to different management and supervisory roles and styles, and empowering employees. This emphasizes the increasing awareness of distinct differences between management and leadership. As stated by John Kotter, "Leadership complements management; it doesn't replace it."[18] Management is needed for order and consistency; leadership is needed to cope with change.

Three qualities that followers want from leaders are direction, trust, and hope, according to Warren Bennis[19]. Basic ingredients to provide these qualities include:

- Guiding vision: Leaders must have a clear idea of where the organization is going and why, and the strength to continue in spite of resistance and failures. Setting direction for the utility creates a specific target, toward which everyone can aim their efforts.
- Passion: Leaders love what they do and love doing it. Communicating this gives hope and inspiration to others.

---

[17] Kotter, *Leading Change.*

[18] Kotter, John P. *John P. Kotter on What Leaders Really Do.* Boston, MA: Harvard Business School Press. 1999.

[19] Bennis, Warren. *On Becoming a Leader.* Reading, MA: Addison-Wesley Publishing Co. 1989.

- Integrity: Trust is derived from integrity, and must be earned. Leaders need to develop self-awareness and deal with their own strengths as well as failings. Covert honesty is overtly exhibited and inspires similar response in others.
- Curiosity and daring: Leaders are eager to try new methods, listen to new ideas, and willing to take risks. They learn from mistakes and motivate others to step out of their comfort zone in trying new ways of doing things.

Kotter differentiates management and leadership by defining management as a set of processes that keep a complicated system of people and technology running smoothly. In his view, the most important aspects of *management* include planning, budgeting, organizing, staffing, controlling, and problem solving. *Leadership* is a set of processes that create organizations in the first place or adapt them to significantly changing circumstances. [20] From Kotter's perspective, leadership:

- Defines what the future should look like
- Aligns people with that future vision
- Inspires them to make the vision happen despite the obstacle

Kotter believes successful organization transformation is 70 to 90 percent leadership and only 10 to 30 percent management. His view on management versus leadership is illustrated in Table 4.1.

---

[20] Kotter, *What Leaders Really Do.*

Table 4.1
Management vs. leadership

Management	Leadership
Planning and Budgeting: Establishing detailed steps and timetables for achieving needed results, then allocating the resources necessary to make it happen	*Establishing Direction:* Developing a vision of the future—often the distant future—and strategies for producing the changes needed to achieve that vision
Organizing and Staffing: Establishing some structure for accomplishing plan requirements, staffing that structure with individuals, delegating responsibility and authority for carrying out the plan, providing policies and procedures to help guide people, and creating methods or systems to monitor implementation	*Aligning People:* Communicating direction in words and deeds to all those whose cooperation may be needed so as to influence the creation of teams and coalitions that understand the vision and strategies and that accept their validity
Controlling and Problem Solving: Monitoring results, identifying deviations from plan, then planning and organizing to solve these problems	*Motivating and Inspiring:* Energizing people to overcome major political, bureaucratic, and resource barriers to change by satisfying basic, but often unfulfilled, human needs
Produces:	*Produces:*
Produces a degree of predictability and order and has the potential to consistently produce the short-term results expected by various stakeholders (e.g. for customers, always being on time; for the Council Members, being on budget)	Produces change, often to a dramatic degree, and has the potential to produce extremely useful change (e.g. new services that customers want, new approaches to labor relations that help make an organization more competitive)

Source: Kotter, *What Leaders Really Do* 1999.

Positive environments cultivate new leadership. Through patient coaching and mentoring, leadership can be developed in all levels of the utility. Identifying and encouraging potential leaders among lower level managers and supervisors demonstrates a proactive approach to traditional command and control cultures, and sends a clear message of looking to the future. Margaret J. Wheatley, renowned author and prominent speaker on organization change issues, recently participated in the Fourth Annual Worldwide "Lessons in Leadership Series: Leadership from the Ground Up: Teaming Up for the Year 2000." Her message focused on relationships within organizations, and that people have a natural need to be together and to help others. Life itself is "system-seeking"; people need to be connected to others and are continuously reaching out in new ways toward creating a shared intent. Leaders are urged to recognize the potentials created by encouraging and empowering shared perspectives that lead to greater creativity in problem-solving and solutions.[21]

Moving decision-making and problem-solving responsibilities toward shared accountability builds and supports relationships. It also enhances individual personal power within a shared setting and promotes consensus-based decisions versus giving orders. In

---

[21] Wheatley, Margaret J. *Lessons in Leadership Series: Leadership from the Ground Up,* 2000.

addition to resulting in more effective decisions, there is buy-in by staff and a sense of inclusion in the leadership process.

CREATE ORGANIZATION ALIGNMENT AROUND A COMMON VISION

Alignment of the organization is the degree to which the entire organization is aligned with a common vision and mission and has embraced a common set of values, principles, and goals. Employees must understand their assigned work processes and practices in the context of delivering quality products and services to customers, as defined in the utility's vision and mission.

For a business to be successful it must, "align the four essential elements that together create growth and profits: strategy, customers, people, and processes".[22] It is only through an approach that aligns all four elements that organizations achieve long-term success.
Vertical alignment brings together the strategy of the organization and its people, allowing them to take effective action. Likewise, horizontal alignment is about the connection between customers and the organization's processes.

Alignment is a set of actions; a skill set that enables leaders to:

- Connect their employees' behavior to the vision and mission of the organization, turning intentions into actions
- Link teams and processes to the changing needs of customers
- Shape business strategy with real-time information from customers
- Create a culture in which these elements all work together seamlessly

Successful organizations proactively create a self-aligning and self-sustaining culture in their organizations by:

- Carefully crafting and articulating the "essence" of their business
- Defining a *few* critical strategic goals and imperatives and deploying them throughout their organizations
- Linking performance measures and metrics to goals, and to a system of rewards and recognition

Aligning a division or an entire organization is an ongoing balancing act that involves setting direction, linking processes and systems, and making constant adjustments. Adjustment is critical to avoid drifting from the target (the vision). Alignment is really the process of continual improvement.

Transition is the hardest part of change projects. The change itself is situational – a tangible situation that is real. Transition is a psychological process that people go through to get to the new state or practice. Unless transition occurs, change will not work.[23] Part of the procedure of transition is leaving old ways and practices behind, and moving on to the new. The uncomfortable time between leaving the old and embracing or becoming engaged in the new, is called the "neutral zone." It is a period when people can be anxious, resistant, and confused. It

---

[22] Labovitz, George, and Victor Rosansky. *The Power of Alignment: How Great Companies Stay Centered and Accomplish Extraordinary Things*. New York: John Wiley and Sons, Inc. 1997.
[23] Bridges, William. *Managing Transitions: Making the Most of Change*. Reading, MA: Addison-Wesley Publishing Company. 1991.

can also be exciting and rewarding, but it is important to recognize, understand, and align this psychological process with the organization's vision and mission to ensure a successful change effort.

CREATE COMMITMENT BY CONTINUOUS PARTICIPATION

Management Commitment and Participation

No amount of new initiatives or strategies will be effective without commitment, support and forthright "showing, not telling," from management. *The Dance of Change* describes what happens when processes that reinforce change and those that limit change collide and conflict.[24] Three reinforcing processes needed to sustain profound change are: enhancing personal results, developing networks of committed people, and improving business results. Some of the challenges to implementing these processes include:

- Management clarity and consistency (walk the talk): Walking the talk includes clear and believable behaviors, inspired by matching underlying values and goals. Visible support by management is crucial to successful change, as is consistent behavior. Even one verbal explosion can demonstrate that autocratic styles are still there and create the suspicion for employees that all of the talk about teamwork is really only talk.
- Control over people's time: Performing new tasks or doing old tasks in a new way takes more time at first. Succeeding with organization change requires time for practice and reflection. Continuous support and encouragement, as well as providing the extra time needed for focus and concentration is necessary.
- Fear and anxiety: Management must set an example of openness, and remind employees that fear of change is natural and to be expected. A safe environment is one that inspires trust and relieves anxiety by acknowledging and accepting different views, skills, and personal styles.

Participation by Working in Teams

"Teams and performance are an unbeatable combination,"[25]. Unfortunately, organizations sometimes forget that teams are only the means, and not the end to success – performance is the primary objective. Key lessons learned and explained in *Wisdom of Teams*, include:

1. "Significant performance challenges energize teams regardless of where they are in an organization." When meaningful performance is the shared goal, groups of people can develop new skills, surmount hierarchical barriers and turf issues, and accomplish significantly more than people working alone.
2. "Organizational leaders can foster team performance best by building a strong performance ethic rather than by establishing a team-promoting environment alone."

---

[24] Senge, Peter, Art Kleiner, Charlotte Roberts, Richard Ross, George Roth, and Bryan Smith. *The Dance of Change*. London, England: Doubleday/Currency. 1999.
[25] Katzenbach, Jon R., and Douglas K. Smith. *The Wisdom of Teams*. New York: McKinsey & Company, Inc. 1993.

The term "teams" must not be confused with "teamwork." Real teams will be successful when leaders look to performance and results rather than focusing on the mechanism with which to meet them.

3. "Biases toward individualism exist but need not get in the way of team performance." Individuals can achieve personal growth and benefits while contributing to the collective strength of the team.
4. "Discipline – both within the team and across the organization – creates the conditions for team performance." Working together to establish, develop and achieve common performance goals, and honing the skills needed to do this, transforms a working group into a team, and requires "disciplined action."

DELEGATE DECISION-MAKING AND CREATE ACCOUNTABILITY

Decision-making in utility organizations has traditionally been the responsibility of management only. Strategies for decision-making in changing organizations include delegating decision-making authority to the lowest level possible. This requires commitment of both management and employees to let go of long-held behaviors and move toward shared accountability and new ways of making decisions. Delegating the authority to make decisions and communicating the freedom to act must be clearly and repeatedly demonstrated throughout the organization. This process takes time for people to accept and act on new responsibilities.

In the Harvard Business Review classic (HBR 1973) the Tannenbaum and Schmidt continuum of leadership behavior illustrates the movement from management-based decision-making to participative decision-making. "If direction is in order, [the successful leader] is able to direct; if considerable participative freedom is called for, he is able to provide such freedom."[26]

The continuum was first published in 1958, and subsequent social pressures and new concepts of management inspired the authors to recognize "relevant interdependencies between the organization and the environment." Their new behavior continuum (shown below) illustrates that the "area of freedom" that is shared by managers and non-managers is "constantly redefined by interactions between them and the forces in the environment." Also, the terms "manager" and "subordinate," were changed to "manager" and "non-manager," to make the difference functional rather than hierarchical.

Other factors that can influence situations in which decision-making is delegated include the amount of time available; the amount of information known and unknown; whether acceptance by others is necessary; whether goal alignment with others is important; and how critical it is to have commitment from others.

---

[26] Tannenbaum, Robert, and Warren H. Schmidt. *Harvard Business Review (HBR) Classic*, Number 73311. Boston, MA: Harvard College. May-June 1973.

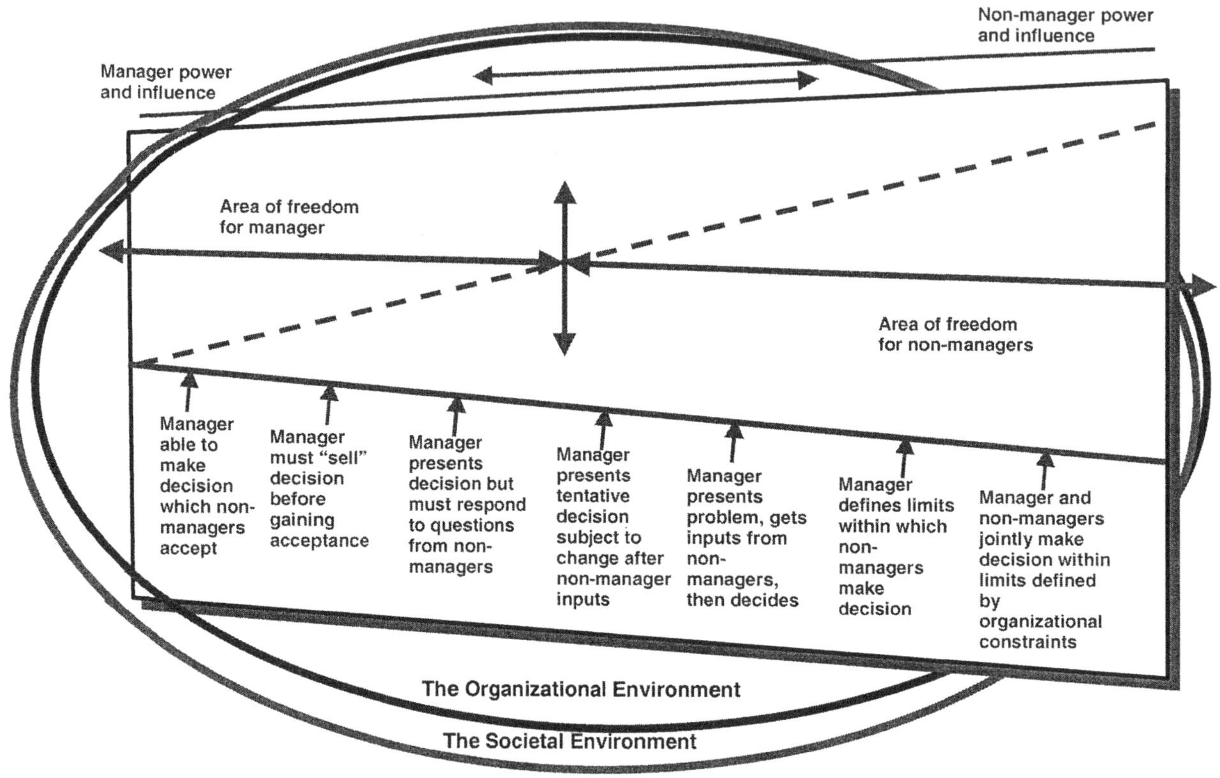

Figure 4.1 Continuum of Manager – Non-Manager Behavior (Source: Tannenbaum and Schmidt, 1973)

In a traditional manager/subordinate organizational culture, employees are used to having decisions made for them. It becomes easier and easier to fall "below the line," in being accountable for actions, and people can fall into a "victim mode," as explained in *The Oz Principle*.[27] In the victim mode, people avoid responsibility by placing blame on other people, circumstances, and the like. Accountability is defined as a process of "seeing it, owning it, solving it, and doing it."

The interdependence of a team environment and shared accountability helps to move individuals "above the line" toward accountability. Leading an organization to shared and responsible decision-making takes time, commitment, and willingness of managers to let go of control.

[27] Connors, Roger, Tom Smith, and Craig Hickman. *The Oz Principle: Getting Results Through Individual and Organizational Accountability*. Paramus, NJ: Prentice Hall Press. 1994.

EFFECTIVELY COMMUNICATE CONTINUOUSLY

Motivating employees to embrace new programs and activities is the anticipated goal of communications efforts related to organization change. Communications involves *effectively* communicating the utility's competitive challenges and their commitment, vision, mission, goals and progress in achieving them.

Creating a communications plan is a key component of change efforts for utilities, and it is necessary to document issues including message clarity, frequency, and repetitiveness, methods of delivery, and a specific plan to transmit the message(s) (how, who, where, when).

Unfortunately, many organizations spend huge amounts of time and effort on developing a vision and mission, and some develop formal communications plans, but fall short in verifying understanding and acceptance of their message. "The real power of a vision is unleashed only when most of those involved in an enterprise or activity have a common understanding of its goals and direction."[28]

> **Kotter stresses communicating organization vision using these factors:**
>
> - Simplicity, no jargon or techno-babble
> - Using metaphors, analogies, and examples—draw a mental picture
> - Multiple forums, such as formal and informal interaction, memos, big and small meetings
> - Repetition, repetition, repetition
> - Leadership by example—behavior that is consistent with the vision
> - Explanation of seeming inconsistencies
> - Give and take—two-way communication

There is a dynamic process of communication tension in organizations. This tension occurs between employees and their perception of reality, and their attempts to modify existing culture to make it more meaningful to him or her. People have their own goals, desires, and incentives, and they will act in ways that are consistent with what they perceive is in their self-interest. It is important for organization leaders to recognize the impact of messages on this tension, to assure that communications are received in the way they are intended. This requires that communications be continually evaluated for *effective* understanding and that people are making sense out of the actions and messages of others.

Communications plans must advocate simple and direct messages and transmitting them in a variety of ways. Easily overlooked is the reality that letters, memos, posters, and even e-mail are passive communication. Active, meaningful, two-way conversation should be a significant element of every communications plan.

Another sector that cannot be overlooked in communicating change messages, is stakeholders external to the utility. Delivering appropriate messages to the media, governing bodies, related utilities such as gas and electric, and others, is important in maintaining credibility and trustworthiness for the utility, as well as support for the change efforts.

ESTABLISH AND TRACK MEANINGFUL MEASURES

Measurement of utility performance is integral to any change process, if the results of new workforce practices are to be quantified. Performance measurement has generated

---

[28] Kotter, *Leading Change.*

enormous interest when perceived as a simple solution to becoming competitive in the eyes of many managers. It has generated great skepticism in the eyes of many utility employees, when viewed as a threat to pay increases, career advancement or even basic employment. Therefore, very few water utilities today practice meaningful performance measurement, where measures are useful to employees in adjusting business performance and seeing the results of their efforts. From other industries and businesses, the definition and use of meaningful measures has been studied and analyzed, resulting in a summary of best practice themes[29]:

- Measures are strongly linked, aligned and integrated with key business initiatives and drivers
  - Strong, clear alignment with organization's strategic planning process
  - Integrated with quality initiatives (ISO 9000, 14000), award programs (Baldrige), or performance improvement programs/strategy
  - Visible leadership through senior "champions"
  - Ties to incentive programs, pay for performance, or at-risk compensation, especially at the executive and senior levels
- Measures are based on a framework such as a Balanced Scorecard[30] or family of measures
  - Vital, small number of measures
  - Financial measures are not the majority
  - BSC used at corporate/enterprise level (typically 10-20 measures)
  - Family of measures (typically 5-10 measures) used at business process levels throughout the hierarchy
  - Usually includes predictive measures for "look ahead"
  - Sometimes includes direct measures of customer and employee satisfaction
- Wide-ranging education efforts are needed to develop and apply measures in operating units
  - Employee involvement in developing operational-level measures, based on business processes
  - Employees can relate measures directly to their jobs
  - Corporate function drives education and implementation throughout all operating units
  - Frequent review of measures with interpretation of results
  - Periodic refinement of measures and "raising the bar" (annual adjustments, target setting)
- Corporate-wide measurement is implemented by systematic data collection with complex "roll up"
  - Regular/periodic data collection through various information sources and systems
  - Employ a variety of technologies, most "homegrown" solutions
  - Formal, complex processes to "roll-up" local data into corporate measurement systems
  - Automation of data gathering needs further development

---

[29] American Productivity & Quality Center (APQC) International Benchmarking Clearinghouse. *Achieving Organizational Excellence Through the Performance Measurement System. Consortium Benchmarking Study*, 1999.

[30] Kaplan, Robert S. and David P. Norton. *The Balanced Scorecard*. Harvard Business School Press. 1996.

  - Tracking of performance over time to show trends and relationships (historical data)
- Widely communicated measurement results are used to involve staff for appropriate business adjustments
  - Formal, regular publishing of results via company news, intranets, etc.
  - Informal, regular, "in your face" communication of measurement results via emails, team meetings, etc.
  - "Roadshow" or high visibility communication by senior leaders
  - Cause and effect relationships analyzed for business process improvements
  - Team involvement and responsibility in adjusting performance

DEVELOP A FLEXIBLE WORKFORCE

Workforce flexibility is a critical element in facilitating utilities to become more competitive and achieve superior customer satisfaction. Integrating efforts and streamlining cooperation between departments and functional operations is key. Other components include:

- Redesigning positions and defining job descriptions
- Succession planning
- Reskilling and competency documentation and training

The following processes are suggested elements of a successful multi-skilled worker program:

- Groundwork: Identify all stakeholders and plan their involvement. Determine early how far the organization is willing to go and the level of complexity it can support. Decide whether full implementation should be done or a pilot project.
- Develop a framework: Determine which skills categories and skill levels exist currently, and which will be needed for the flexible worker. Also determine which skills other than technical will be included, i.e., soft skills for working better together, such as conflict management, problem solving, and communications.
- Workload analysis and skills audit: Analyze tasks performed, collect data from work orders, job descriptions, interviews, focus groups, observation, and other resources. Plug the tasks into skills categories and levels.
- Certification requirements and process: Decide what will be required for initial certification in each skills category and level, and also to maintain certification, licenses, and other requirements. Determine how employees will demonstrate their competence in skills categories and levels.
- Training: Determine what types of training are needed; technical skills, people skills including diversity training, and team training. Determine what resources will be used to train employees; on-the-job with peer trainers, internal training courses, computer-based training, or outside seminars. Some utilities partner with an outside academic facility.
- Collective bargaining and other issues: Identify what constraints may be imposed by labor agreements, personnel policies, and other regulations.
- Compensation: Compensation is directly linked to productivity and the multiple-skilled worker. A workforce that demonstrates a broader skills-set and can perform a wider variety of tasks allows utilities to perform the work with fewer people. When

people are more productive and bring additional value to the utility, they deserve to be compensated at a rate commensurate with their contribution to the utility. This requires a flatter and broader pay structure.

PROVIDE APPROPRIATE COMPENSATION AND REWARDS

In any significant change initiative, the areas included under the compensation and reward umbrella must be reassessed to insure that the new behaviors desired by the organization are reinforced. The organization should begin by determining the objectives of the compensation system (such as recruitment, retention, retraining) as they relate to the strategic plan of the organization. Once the objectives have been determined, the desired behaviors can be identified. After the objectives and behaviors have been established, a compensation structure can be developed that supports the organization's strategic direction.

Developing the appropriate compensation structure is a critical step in moving the organization change past process to people. "Pay is first and foremost a people issue".[31] It is about motivation and reshaping and refocusing of behaviors.

A total compensation plan may include individual, team, and organization-wide levels of incentives. Some types of individual pay systems that could be included in the compensation framework are:

- Pay for performance: compensation for completion of work as measured against defined goals and objectives.
- Pay for knowledge and skills: compensation for skills required to perform any number of functions rather than for a specific job.
- Pay for competencies: compensation for interpersonal traits and abilities such as working in teams, decision-making, and customer service approach.

In addition to compensation, there are other benefit and rewards systems that can be used to support the organization's vision, values, and business strategies. These can include recognition events, access to various employee services, and organization-sponsored employee activities. Guidelines regarding issuance of these benefits and rewards need to be included in the compensation/reward structure.

To be effective, compensation and rewards must also be integrated with other human resources processes, such as training and performance management. Continuity of messages regarding human resources processes is one of the most direct ways that an organization has of communicating with its employees.

"There is no one right total compensation mix for all organizations".[32] Every organization must consider its own individual culture in developing its compensation and rewards structure.

USE SUPPORTING HUMAN RESOURCES TOOLS

Building the information-based knowledge sharing work environment that is required in a changing organization means learning to think of data and systems in new ways.

---

[31] Hofrichter, Platten. 1996. [missing reference]
[32] Lawler, Edward E., III. *Strategic Pay: Aligning Organizational Strategies and Pay Systems.* 1990.

Human resource systems must provide tools that support routine, yet critical, administrative processes while enabling individuals to focus time and energy on strategies that support the entire organization. By helping increase data accuracy and reduce unnecessary work, an organization's Human Resource Management System (HRMS) can increase service levels, access to information, and improved effectiveness. A truly effective system should provide the right tools to the right users throughout the organization.

The major categories of information in a HR database (competency management, time and attendance, training, recruiting, compensation and benefits, and payroll) should be integrated with an organization's other information databases.

The difficulty in successfully utilizing an integrated HRMS is the gap between technology and culture. Emerging systems allow managers to process and view data in new and interesting ways, and they also allow organizations to understand the true costs and real value of people and processes. These systems are now forcing individuals to share information in environments where data exclusivity has been the norm. Managers need to be aware of how the data they have responsibility for is used throughout the organization and be willing and able to respond to the new organization's increasing demands for information.

Beyond the flow of data throughout the organization, information sharing must also include feedback on individual behaviors. Systems such as 360-degree performance feedback provide information from several different perspectives and allow an individual to adjust behaviors as appropriate.

The feedback mechanisms need to provide for real-time information and learning opportunities. Feedback should be provided that is constructive, specific, and non-anonymous and includes suggestions for improvement or encouragement to continue the desired behaviors. Guiding principles of On-The-Level Communication[33], regarding performance feedback include directness, respect, shared responsibility, and purpose. All individuals in the organization should receive guidelines on giving and receiving feedback in areas that support the changing environment.

Organizations that understand the advantages gained by the effective use of HR tools build systems, both technological and cultural, that will support the changing environment.

BUILD AND MAINTAIN EFFECTIVE RELATIONS (WITH GOVERNANCE AND LABOR)

In organization change efforts where governance and labor are not actively participating, up to 70 percent will fail to achieve the desired results.[34] People are resistors to change and the stakeholder entities of governance and labor represent potential sources of significant resistance. Keys to successfully involving these stakeholders include:

- Identifying their motivators: What is important to them? What do they perceive as success? What do they want to avoid?
- Identifying their enablers: Who are the individuals with the personal and positional power to influence the change process (positively and negatively)?
- Identifying their triggers: What are the conditions that link the motivators to the enablers?

---

[33] McLagan, Patricia, and Peter Krembs. *On-the-Level: Performance Communication that Works.* 1995.
[34] Scott-Morgan, Peter. *The Unwritten Rules of the Game: master Them, Shatter them, and Break Through the Barriers to Organizational Change. 1994.*

When using this approach, it is important to identify what people perceive the motivators, enablers, and triggers to be as well as reviewing those stated by the organization (the two may not be the same). This allows an organization to determine some of the behavioral barriers to change efforts.

One area of significant behavioral barriers in change efforts is within the governance structure itself. "Framing the governance challenge more effectively can go far beyond merely eliminating common problems; it can provide a clearing for strategic leadership".[35] John Carver's Policy Governance model is being utilized in nonprofit and public organizations to help organizations successfully navigate the new ways of thinking and operating necessary in the competitive environment. This model groups policies into the following categories:

- Ends: What human needs are to be met, for whom and at what cost or relative worth.
- Limitations: Those principles of prudence and ethics, which limit the choice of means (practices, activities, circumstances, methods).
- Relationships: The manner in which power is passed throughout an organization.
- Process: The manner in which "ownership" is represented and strategic leadership is provided.

Understanding of the ends, limitations, relationships and process of governance and labor provides a solid foundation for building and maintaining the relationships necessary for a successful change initiative.

[35] Carver, John. *Boards That Make a Difference: A New Design for Leadership in Nonprofit and Public Organizations*. 1997.

CHAPTER 5
WATER UTILITY "BEST PRACTICE" CASE STUDIES

Examples of successful practices from four water utilities are described in this Chapter. These practices are examples of how various utilities have employed the strategies for organization change in different ways. Information was collected through interviews of utility staff, and is summarized here. Full Data Collection Forms are found in Appendix B.

The practices described are typically part of a larger initiative the utility has embarked on. Each initiative employs one or more of the strategies for organization change (described previously in Chapter 4). Interdependence of strategies is a significant factor in each of the practices described. During the interview process, it was noted that for every initiative, there were resulting interactions with other areas as the practices progressed. For example, during Ottawa-Carleton's water distribution system excavation process, which is shown below as a workforce flexibility practice, there was resulting transfer of staff and significant cross-training which drove the creation of a full-time training coordinator position, and a computer-based training program. In addition, leadership played a crucial role in supporting the practice and high-level management involvement was critical to the success of the program.

Therefore, each utility initiative includes practices that address more than one strategy for organizational change. Table 5.1 highlights the most relevant areas involved for each of the successful practices.

Table 5.1

Example practices from strategies for organization change

UTILITY:	Portland Water Bureau	Portland Water Bureau	Seattle Public Utilities	Ottawa-Carleton Water Division	Louisville Water	Louisville Water	Louisville Water
TITLE OF INITIATIVE:	Knowledge Transfer	Utility Worker Classification	Organization Performance System	Work Process Redesign	Union Relations	Performance Recognition	Working in Teams
STRATEGIES:							
Alignment		✓	✓		✓		✓
Urgency			✓				
Leadership			✓	✓	✓		✓
Decision-Making			✓				
Participation	✓	✓	✓		✓		
Communications	✓	✓	✓		✓		
Measures			✓	✓	✓		
Workforce Flexibility	✓	✓	✓	✓			✓
Compensation			✓			✓	
Tools		✓	✓	✓			
Relations		✓	✓		✓		

ALIGNMENT OF AN ORGANIZATIONAL PERFORMANCE SYSTEM: SEATTLE PUBLIC UTILITIES

As a result of a re-organization in 1997, Seattle (WA) Public Utilities was created to provide integrated utility services to Seattle citizens. The new utility provided opportunities to rebuild, improve and integrate existing business and human systems.
Seattle Public Utilities has 1200 employees and a $500 million operating budget, with a ten-year capital program exceeding $1 billion. The organization consists of:

- Water Supply
- Drainage
- Wastewater
- Solid Waste
- Engineering Services

Visionary leadership and a mayor who is committed to measurable results created a formula for success. This formula provided a commitment of resources and money to build the Utilities' Organizational Performance Framework.

The program framework is in various stages of development and implementation, having begun when the Utilities formed in 1997. The framework aligns the work of the Utilities with its Vision and Strategic Goals, and consists of four components:

1. **Strategic Business Planning**: Setting and defining vision, mission, values, goals, and results in a 2-year strategic business plan.
2. **Operational Planning**: Develop work unit performance measures and targets, operational work plans, and a budget.
3. **Performance Measurement**: Collect feedback (customer, financial organizational, operational), measure and evaluate. Products include a performance measurement database, customer feedback system, and a reporting system.
4. **Performance Management**: Use data to manage change, make decisions, reallocate resources. Create/adapt business systems and continually develop workforce skills, knowledge, and leadership. Products include utility-wide leadership core competencies, 360-degree management evaluations, employee surveys, leadership and employee performance and development planning, management performance contracts and employee performance reviews linked to performance measures.

The four components are further described below:

1. Strategic Business Planning

The first strategic planning process conducted in early 1997 was a top-down, bottom-up approach and produced a two-year (1997-1998) Strategic Business Plan. Over 250 employees from department head to line worker were involved in creating the Utilities' vision, mission, goals and objectives. Employees identified 50 strategic efforts that were linked to accomplishing the objectives and were published in the Plan. At the end of 1998, the Strategic Effort Result Report identified that over 95 percent of the efforts had met the targeted completion dates.

The second strategic planning process was another top-down and bottom-up approach. The nine-member executive team set the strategic direction for the department, reviewing the

1997-1998 Strategic Goals, developing result statements and aligning the Utilities' Strategic Business Plan with the mayor's "Managing for Results" program—G.A.R.M. (Goals, Actions, Results and Measures). The bottom-up approach was focused on involving employees at the work unit level in the development of performance measures and connecting their everyday work with the strategic goals of the Utilities.

To involve employees, a Strategic Business Plan campaign with a performance-focused message was kicked-off at a March, 1999 Utilities Leadership Forum with over 150 Utilities management staff – from department director to line crew supervisor. The mayor attended as guest speaker to talk about how the work of the Utilities supported the goals of the city and what the Managing for Results program was all about. The Utilities Director reinforced the important messages about performance and results, using the Utilities' 1999-2000 Strategic Business Plan to make the connection.

A series of five supervisor workshops was conducted in March 1999 to further educate line supervisors about the Strategic Business Plan and prepare them to serve as key communicators of the Strategic Business Plan to their employees and help them develop meaningful performance measures for their unit's work. The workshops provided the supervisors with tools to better understand each of the six strategic goals and how they could help employees connect their everyday work to the Utilities' strategic goals. The supervisors received baseball hats in recognition of their important role of communicating the Plan to employees.

During the month of April, supervisors conducted staff meetings with their employees to roll out the Strategic Business Plan. Every employee was asked to demonstrate how his or her work connected to a strategic goal. Each employee received a copy of the Plan and a Strategic Goals Note pad.

A monthly news article in the *SPU News* features an employee interview about how his or her work connects to a strategic goal of the Utilities. An internal web site for the Utilities' Strategic Business Plan is available to employees and will link to the operational work unit performance measures at a future date.

2. Operational Planning

In February, 1998 the operational planning process began. The existing Utilities 1998 Work Program served as the starting point to develop the G.A.R.M. report. The goals, actions, and results were revised through the Strategic Business Planning process.

3. Performance Measurement

The process of developing performance measures began in summer 1998. An operational work plan format was developed in alignment with the required G.A.R.M. report.

A two-day Utilities workshop with the 39-member leadership team, senior budget and policy staff (approximately 30) and appointed performance deputies (6) was held in October 1998, to conduct performance measurement training, and begin development of unit performance measures using budget activities.

Over the next three months, work units developed 1999-2000 operational work plans that included performance measures and initiatives. Work units set targets, identified baselines (where possible), established data sources and assigned data managers for each of the performance measures. Executive Team reviews occurred at regular intervals with the final operational work plans approved in January 1999.

In 2000, the Utilities focused on acquiring customer feedback and look into the benchmarking and best practices of other utilities.

4. Performance Management

Performance management includes defining the Utilities' culture, developing leadership to manage change, and recognizing, rewarding and developing employees. It is the "human side" of performance.
Several projects have served to set the stage to move forward in the performance management process:

- In January 1999, the Utilities Executive Team established a set of leadership core competencies and conducted a pilot 360-degree evaluation with its 39-member leadership team.
- In July 1999, an employee opinion survey, the "Employee Voice," was conducted to set a baseline of employee satisfaction. The items were based on the Organizational Values and Work Principles.
- In 1997, performance contracts for senior management and key policy staff were developed that aligned with the Strategic Business Plan. In 1998, the contract system was refined to include performance measures.
- In 1999, the employee performance review system was revised to align with the Strategic Business Plan, include the leadership core competencies for supervisors and evaluate the knowledge and usage of performance measures.

Lessons learned from Seattle's Performance System:

- People must be willing to take ownership and use the system
- The system must add value
- Constant communication is critical
- The change process is continuous
- Build on what you already have
- Be flexible—it doesn't have to be perfect the first time around
- Watch what you measure because you may get an unexpected result
- If the system is perceived to be too cumbersome, it will not be supported

The performance measures database went into design in May 1999. Data input began in August with completion in October 1999. Training of department staff on the database will occur in the last quarter of 1999. It is expected the data managers or a designee will derive the performance data for the 12-month actuals from the database.

Performance sessions were conducted with the Department Director during July 1999. Six-month actuals were presented, targets and performance measures were discussed and refined over the next several weeks. Performance sessions, using the operational work plans, will occur in bi-annually with the Department Director to discuss performance, targets, needed adjustments/refinements.

The Utilities G.A.R.M. was reported in July 1999 to the City Council and Mayor. The G.A.R.M. will be reported bi-annually. Additional reporting will occur at 12, 18 and 24-month intervals.

In 2000, an SPU customer scorecard was designed and distributed to interested stakeholders.

PARTICIPATION – WORKING IN TEAMS: LOUISVILLE WATER COMPANY

Results of Louisville's practice:

- Higher quality work life
- Employees feel they are making meaningful contributions
- People have grown professionally and are more involved in their work
- Open communication and shared decision-making

In 1992, the Louisville (KY) Water Company (LWC) made the decision to become a team-based organization, as a result of new leadership in the Company and an overall organizational change effort. This effort required a significant commitment to establishing expectations and goals, and providing the appropriate tools and training.

An outside consultant was hired to provide initial training to three cross-functional project teams, starting with the operational areas. The Company President, John Huber, provided input and involvement, and "drove" the process. The consultant helped the teams learn about roles and responsibilities, and how to conduct and lead meetings. They provided a learning process for managers to let go and encourage teams to work toward becoming self-directed.

The team concept is now well established and fully implemented throughout the Company. Natural Work Teams (NWTs) are part of the LWC systems-based management structure. Business System Owners and Process Owners along with the NWTs practice systems-based management and are accountable for the delivery of products and services.

LWC has moved from a traditional hierarchical-based organization to one where NWTs set and achieve common goals, and working as teams has become embedded in the culture of the organization. This includes long-term work teams, as well as shorter-term project related teams. Another program, the "Facilitators Services Program," started as a result of the teams process. There are 22 trained facilitators who spend 15 percent of their time facilitating teams.

The Human Resources Department assumed the responsibility for training from the consultant, and currently trains all new hires. Metrics have been established to measure progress of teams, and are continuing to be developed. Over the past three years, everyone in the organization has received team training. LWC has a corporate scoreboard or a "corporate family of measures" that was developed internally, and is still evolving, based on key results areas. The number of employees involved in teams is tracked, and the process improvements that teams work on, i.e., what was produced, what worked, what happened to the results, who chartered the team, and who removed the barriers so that the team could be more effective.

Work is still being done on measuring dollar savings, and measurements will be integrated in order to make decisions based on the data analysis.

There was resistance from some – especially managers – it was a learning process for them to let go in order for teams to really become self-directed. Time and training helped managers to support the practice.

According to Beverly Irons, former Director of Workforce Diversity, "The results were not especially surprising – we expected to be successful. But, it was gratifying to really see that it happened... The organization culture has changed and even if the project driver were no longer there, we would not revert."

KNOWLEDGE TRANSFER: PORTLAND WATER BUREAU

The City of Portland (OR) Water Bureau wanted to have a systematic approach to workforce planning. The utility has a large number of people nearing retirement and needed to plan for filling vacancies. They also wanted to redefine positions in order to create a more flexible and diverse workforce with the level of skills needed to do the work. As a result of the work force planning project it was recognized that a more formalized process needed to be in place in order to capture and retain information that might leave as individuals retire from the utility.

The practice involves transferring knowledge from a large number of potential retirees to the newer workforce so that institutionalized personal knowledge isn't lost when people retire. Each work unit is working toward capturing personal information, experience, and knowledge so that it becomes part of the organization. Some of the methods and processes used to support knowledge transfer include:

- New employees are being hired before the person they are replacing leaves, so that training and knowledge transfer can occur as new people learn skills from their predecessors and work into their positions.
- Retirees are being rehired on a part time basis to train new employees.
- Videos are being made of people doing their work, and work units are documenting personal information, experience, and knowledge so that it becomes part of the organization.
- Procedural manuals, standardized record keeping formats, archiving and data retrieval systems are being designed and implemented.
- Cross-training of existing staff is occurring.
- A new GIS (Geographic Information System) is coming on line that will help to support this practice.
- A new CIS (Customer Information System) is being implemented that will allow the utility to capture additional information about specific meters and addresses.

WORK PROCESS REDESIGN – WATER DISTRIBUTION SYSTEM EXCAVATION: OTTAWA-CARLETON WATER DIVISION

A sense of urgency led the Regional Municipality of Ottawa-Carleton (CN) Water Division to design and implement a process for their water distribution system excavation. The practice includes all aspects of planned and emergency maintenance when an excavation is required to perform the work.

Based on metric benchmarking (similar to the Xerox method) the utility's excavation process was not considered best in class. An eight-member team was formed to study the process and improve it. The process, championed by high-level management, took six months to complete, and implement, and began December 1998.

The team developed detailed plans that described actions, responsible persons, monitoring and costs. They went on two site visits to see some best-in-class processes in operation, and held debriefing sessions while still on the road.

Recommendations from this process resulted in transfer of staff (Corporate fleet operators transferred to the water division to become "true" team members of the excavation crew, and to become cross trained), cross training and adopting new equipment, i.e., a backhoe with extended reach. Following this practice, management staffed a full time training coordinator. A training room with 3 computers was set up on-site for computer-based training.

Better work planning and reduction of equipment were a direct benefit. Better staff morale, and the ability for staff to do more kinds of work allowed for the reduction in crew size, while providing management with more flexible staff. Staff's involvement in the process led to better accountability and ownership of the process.

Direct cost savings amounted to $210,000/yr, with increased productivity savings of $540,000/yr. These savings take into account the Training Coordinator and the computer training room, but do not take into account the purchase of new backhoes. These will be purchased as the older ones are replaced.

The results of the practice are included in long-term planning, and short-term recommendations are made for training and necessary equipment purchases. A mid-management champion is assigned the responsibility to make certain the practices/recommendations are being implemented over a specific time period. Some ways to measure the benefits of the practice include:

- Reduction in staff numbers (moved from a 7 to a 5-person crew).
- Reduction in material use.
- Reduction in time on-site.
- Actual reduction in budgets and work order cost.

Lessons learned:

- Teams must be representative of the functional areas
- Focus on action
- Team must have ownership of the process
- Teams must meet deadlines
- The process and implementation must have a "champion"
- Selection of benchmarking partners is critical
- Training essential to develop "in-house" expertise
- Management must be committed to implement recommendations
- Financial support for the process must be in place

Long-term data and planned implementation of activity-based costing (ABC) will help the Region monitor these performance measurements. Currently, the water distribution computerized maintenance management system captures most of these costs.

A reduction of staff is imminent when best practices show a reduction of two staff per crew should take place. People fear losing jobs. Due to the age of its workforce, the Region was able to meet this objective through attrition. Initial objections to cross-training were overcome with pilot training programs implemented with crews willing to try something new.

UTILITY WORKER CLASSIFICATION – WATER SERVICE UTILITY WORKER APPRENTICE PROGRAM: PORTLAND WATER BUREAU

The City of Portland (OR) Water Bureau wanted to have a systematic approach to workforce planning, and to create a more flexible and diverse workforce with the level of skills needed to do the work. A program was developed to allow for the hiring of unskilled employees and to bring them to journey level within two years through a highly structured training program. Susan Bailey, the Water Bureau's Program Manager for Workforce Development, was a member of the City's Apprenticeship Task Force (ATF), a group tasked with creating apprentice opportunities within the city. Mort Anoushiravani, a Senior Manager at the Bureau, was instrumental in fostering the Water Bureau's flexible worker position. Susan worked with the ATF, State of Oregon Bureau of Labor and Industries, the City's Trades Apprenticeship Committee, and community based organizations, many of which represent minorities and women who want to get into the trades.

After the beginning concept of flexible workers, a thorough job and task analysis of utility worker positions was done, with the help of a consultant hired through a local community college. The process of seeking and getting approval by the State of Oregon and of creating the new position at the City of Portland took much longer than anticipated. This resulted from organizational changes as well as the fact that the City as a whole was in the middle of a large job re-classification study. This resulted in the project being delayed six months.

Some benefits and results:

- A more diverse pool of applicants
- A systematic approach to workforce planning
- The potential to have trained and flexible workers with the necessary skills to fill needed positions as they occur
- Enhanced recruitment and selection procedures that utilize community resources to create a more diversified applicant pool
- Enhanced position for the City as an employer of choice

The practice creates flexible utility workers via a formal 4000-hour State-approved apprentice program using training and mentoring. This is an official apprenticeship program overseen by the City of Portland's Trades Apprenticeship Joint Labor and Management Committee. The program provides skilled and flexible utility workers. It also enhances the City's labor and management partnering, and has become a "show-case" project for the City, as well. The City's Bureau of Human Resources is working with the Water Bureau and the ATF to enhance recruitment and selection practices using the approach developed for this program as a model.

The process took approximately 24 months from the beginning concept of the program and task analysis, to hiring the first six apprentices as of September 9, 1999.

As the program developed it was discovered that people who complete the apprenticeship process could serve as a "feeder" positions for other areas as skills develop and mature. The Utility Apprentice Program has become the desired entry point for a variety of positions in the Maintenance, Construction and Operations Groups of the Bureau. It was also determined that a program like this needs to be started early on to prepare for anticipated employee retirement. Because this program will be bringing in large numbers of new employees, many of whom have less experience, are younger and from more diverse backgrounds, the Bureau is developing additional programs to help ensure their successful integration into the workforce. Existing journey level employees and supervisors have been trained in on-the-job training and coaching techniques. Training in creating a respectful workplace is being designed for all employees, and a mentoring program for all new or transitioning employees is being developed.

COMPENSATION – PERFORMANCE RECOGNITION: LOUISVILLE WATER COMPANY

Since 1991-92, employees of the Louisville (KY) Water Company (LWC) have established their own performance goals for the year. Employees are evaluated against those goals and given a rating for their wage increases. In response to criticism from employees that there were not enough increases and goals were not standard, LWC restructured and modified percentages for performance recognition. A Work Planning and Performance Management system (WPPM) established a market-driven structure in April 1999.

LWC's Human Resource Business System was responsible for implementing the program. All systems are eligible and all employees can participate.

The program has grades of 1-14, with market value and range as a control point for each grade. The WPPM has developed a quartile system to measure performance and establish a rating for the grades.

Some results of the program include:

- People feel better about their work – they are motivated by the market-driven structure and can accelerate toward market.
- It has increased motivation and helped to retain employees.

In addition to the market-driven performance recognition model, there are three new ways to reward employees. These include:

- Token System: Each year, every employee is given 12 tokens with a value of $2 each. The tokens are to be given on the spot to fellow employees for recognition of exceptional service or notable achievements. The tokens can be accumulated by recipients and redeemed at local restaurants, discount stores, etc.
- Charter System: Each year, each employee can be recognized if they participated in a charter team, i.e., a baseball team, and someone recommends them for recognition. There is a formal chartering process with criteria such as having a sponsor and a facilitator. Vouchers for $25 are given to each member of the team, up to $250 per team.
- Quality Celebration System: Once a year employees can be recognized up to $12 per employee to celebrate what they were able to do to promote quality. Employees must write up what they want to do, i.e., some form of recreation like mini-golf, etc. or a team building exercise, and submit the form to the Quality Office for action.

UNION RELATIONS: LOUISVILLE WATER COMPANY

The Louisville (KY) Water Company's organization-wide systems-based management approach moved them from utility "departments," to a water utility based on systems, processes, and quality. The cornerstone of this approach is a partnership of all employees.

During the late 1980s and early 1990s, LWC and its union had an adversarial relationship. This was evidenced by a high number of grievances and arbitrations. The union couldn't get the work it wanted, and the utility couldn't get the cooperation or efficiency that it wanted.

Realizing that they must form a partnership in order for everyone to be successful, a Decision Partner Team (DPT) comprised of union and company representatives was put in place. Union representatives include the President and Vice President of the union and six other chairs. Recognizing the need to have front-line workers on board, the six union representatives are selected by the union executive board. The company has fewer chairs, and includes the Business System Owner of Human Relations (consisting of the processes of labor and employee relations, compensation and benefits, recruitment, development and retention of employees). It is important to note that the DPT does not cover grievances or arbitration issues, but meets weekly to look at opportunities within the organization to further develop the partnership between the company and the union, and enhance communications efforts.

The union relations practice was started in 1993-94, and is currently in the implementation stage. The program has received several local awards, including one from the Labor Management Center of the University of Louisville.

Results and benefits of the practice include:

- A "Union Quality Coordinator" position was created. This is a person selected by the union who has a chair on the DPT and spends 100 percent of their time looking for opportunities to improve quality.
- There has been a significant reduction in grievances, arbitrations and union issues. Also, membership in the union has increased by 20 percent.
- It was recognized that the union lacked consistent direction. Union members held a retreat to develop their own Vision and Mission as well as a Strategic Plan to support the company's Strategic Plan.
- Measurements were developed; the DPT identified current levels of involvement, and where they wanted to be, and developed a plan for how to get there. Using a scale of 1 – 5, level of involvement is measured quarterly and annually against the original baseline.
- At the start of the program, the union wanted greater participation. Once they got involved through the company's invitation, they found that they really didn't need as much involvement as they thought. Likewise, the company felt rather threatened or undermined at first, because of the union participation. Ultimately, both sides became more comfortable and reached an appropriate level of involvement.

Of course, these case studies are just a sampling of the many successful programs implemented by best practice utilities. They are meant to illustrate real-life examples of workforce change and the positive impacts for utility management, workers and ultimately for the public they serve.

APPENDIX A
DESCRIPTION OF WORKFORCE ELEMENTS

DESCRIPTION OF WORKFORCE ELEMENTS

Workforce Element	Description
Methods for Assessing Compensation Levels	An organization's compensation system reflects differences in the fundamental market value of the different kinds of work being performed. The compensation structure identifies rules on hiring rates and promotion increases; defines promotion, demotion, and lateral movement; and specifies the timing of and reasons for pay increases. There are two ways of assessing compensation levels. They are: ◆ Direct market pricing of jobs: Useful if there are few jobs. Consists of finding salary surveys that provide information on the jobs and the average rates paid. This average becomes the target rate for a fully experienced job incumbent. ◆ Job grading: Presents a set of core jobs ranked according to the jobs' perceived internal equity. These jobs are market priced and are arrayed in order of their rankings so that a statistical relationship between the rankings and the market price can be derived. A midpoint value is established and each midpoint value is highly correlated with the market value for the core jobs. All other jobs are related in responsibility to the core jobs, and their values are derived from their relationships (higher than or lower than core jobs.)
Union Contracts/Labor Relations	These elements would include proactive/collaborative collective bargaining, third party (binding or non-binding) arbitration, joint labor management teams for process improvement, safety and training, new incentive and skill-based compensations jointly introduced by labor and management, and the creating of multi-tasking job descriptions.
Performance Measurement	Performance measurement is the process of aligning business results, performance processes and products, and performance support. Proper performance measurement techniques achieve two things: 1) Focus on important measurable workplace performance (Timeliness, quality, cost) 2) Focus on complete set of key workplace variables including performance standards, feedback, incentives, goals, guidance, instructions, tools, materials, and the knowledge of skills of the workers.

Workforce Element	Description
Consolidated Position/Job Descriptions	The main reason for creating job descriptions is so that it can be compared to similar work in the labor market. Also, it allows for the relationship between the work and that of other jobs in the organization to be documented. A good job description is written simply, contains a statement of job responsibilities, and differentiates itself from other jobs in the organization. The statement of job responsibilities usually includes the following items: 1) General statement of responsibilities. (i.e., for an accountant – Performs accounting functions according to GAAP.) 2) List of specific responsibilities associated with the job. The list should consist of concrete tasks and allow for the reader or performer to distinguish the job in question from other, related jobs. (i.e., for an accountant – classifying expenses, trial closings, producing statements of income and expense, analyzing accounts, etc.) 3) List of other compensation factors, such as required skills. (i.e., ability to communicate effectively, to solve problems, to organize and coordinate the work of others, to manage conflict, etc.) 4) Working conditions and bona fide occupational qualifications, or the minimum qualifications the candidate usually possesses in order to perform satisfactorily. (i.e., for an accountant possession of CPA)
Innovative Management Techniques	This category could include having union leadership represented on governing boards and management committees, customer involvement in strategic and capital planning, new internal and external communications strategies, and bringing automation tools such as process control and other information management systems to bear on improvement programs.

Workforce Element	Description
Education and Training	Training is "experiences provided primarily by employers for employees, designed to develop new skills, knowledge, and attitudes that are expected to be applied immediately upon (or within a short time after) arrival on or return to the job." (Reynolds 1995, p. 126) Training consists of following four basic interventions: 1) Objectives – Spell out the expected outcome in terms of what the trainee will be able to know or do after the training. The classic training objective contains a performance intent, criteria, and conditions. Usually, the objectives describe new skills or knowledge to be acquired. 2) Content – The information presented to help trainees learn what it is they need to learn. The content is intended to be directly related to the objectives, at the right level for the audience, and sequenced according to skill development needs. 3) Interaction – Opportunity for the trainee to demonstrate what he or she has leaned. If the interaction demonstrates that learning has not taken place, a prescriptive path may be developed to allow additional clarifying content and interaction until learning does take place, a process called remediation. Interaction can take many forms, ranging from simple questioning to sophisticated simulations of on-the-job experiences. 4) Feedback – The information, or comparative data, provided after interactions so that the trainee can compare the interaction to an exemplary model. In the more real interactions, feedback provides a model of performance as it is actually done on the job. Training serves its best purpose when the performance need is to establish new skills and knowledge or to fill a performance gap (to establish skills or knowledge that current performers do not possess but which are needed.)

Workforce Element	Description
Team/Delegated Work Methods	"Team performance is governed by a set of attitudes and actions that help team members to work together cooperatively so that they can effectively achieve the team's goals." (Langdon, p. 365) Teams are appropriate to use in the following situations: when leadership is needed throughout the organization, when diverse viewpoints and knowledge is needed to solve a situation, when processes cross functional areas, when skills are not being used efficiently due to workforce structure. High team performance results when team members work together to achieve a shared objective. In order to achieve this shared objective, every member of the team has his or her own responsibility for particular tasks, but all the tasks are interdependent. Therefore, in order for one person to succeed, everyone must succeed. In order for team to achieve high performance, it must have the support of management. Management is responsible for providing the time and training that teams need to become established, as well as providing feedback and rewards that reinforce teamwork. The obstacle to high performance in teams include: poor management support, uncooperative members, poor team leadership, lack of training, insufficient resources, and poorly articulated goals. The optimal team size is seven members, however a team with five to nine members also performs well. For the team to perform well, the following roles must be filled: ♦ Sponsor: A manager who has set up or empowered the team to perform its tasks. Provides resources and feedback. Must be committed to the team's success. ♦ Facilitator: Team expert that gets the team started and serves as a resource to help members function together effectively as a team. Helps to prevent and solve problems within the team. ♦ Team Leader: Calls and conducts meetings, acts as liaison to management, serves as interface with other teams, keeps team focused, solves problems between team members, deals with disruptive team members, and removes roadblocks to performance. ♦ Team Members: Individuals who are experts in various tasks and activities that must be carried out for the team to achieve its goals. Members must know their jobs and be trained to operate in a team environment.

Workforce Element	Description
Succession Planning	Succession planning is a procedure for identifying people who can replace key executives and professionals and deciding which development programs are needed to prepare them for moving into these higher positions. There are three key elements of an effective succession plan. They are: 1) The succession planning statement – Mission statement that sets out the objectives and top level commitment of the organization to succession planning 2) The succession planning back-up summary – Lists the key positions and the back-up personnel are identified 3) Individual development plans – Individual development plans for each identified successor are made and carried out Succession planning should be done for at least four levels of the organization – the president, vice president, director, and manager. It should eventually be done at all levels, but these four levels are the most important.
Employee Recruitment/Competency	The objective of employee recruitment is to find the most qualified people for the position. Recruitment strategies vary from company to company. Some companies conduct pre-employment screening tests in an attempt to better determine if the candidate is qualified for the job. Competency tests vary from industry to industry.
Workforce Reduction/Redeployment (Outplacement)	When a company finds that some of their employees do not have the skills, abilities or knowledge needed to achieve their assigned goals, they often find themselves in the position of laying off these employees and using outplacement firms to help the employee in their career transition The selected firm typically provides consultation to management, facilitation of meetings with departing employees, and the design and delivery of programs for departing employees.
Workforce Empowerment Techniques	One way to increase employee performance is by empowering employees. Empowering, or motivating, employees can be done in the following three ways: 1) Giving people the ability and the permission to accomplish goals 2) Create positive working environments 3) Place high values on achieving goals.

APPENDIX B
SUCCESSFUL WORK PRACTICES - DATA COLLECTION FORMS

SUCCESSFUL WORK PRACTICES - DATA COLLECTION FORMS

Utility:	**Seattle Public Utilities**
Person Contacted:	**Joanne Peterson, Dove Albert, Cyndy Holtz**
Practice:	**Alignment of an Organizational Performance System**
Describe the practice.	The Utilities' Organizational Performance System is a framework that consists of strategic planning, operational planning, performance measurement, and performance management. This framework aligns the work of the Utility with its Vision and Strategic Goals. **Strategic Planning** sets the direction of the Utility, stating its vision, mission, values and principles, setting strategic goals and defining results. **Operational Planning** connects the operational work plans, which include unit performance measures and initiatives, to the Strategic Business Goals. **Performance Measurement** tracks, monitors, reports, evaluates and assesses Utility performance. **Performance Management** develops and sustains an organizational culture and workforce focused on efficiency and high performance.
How long has it been in place, and what stage is it in (i.e., assess, vision, design, implement)?	The program framework is in various stages of development and implementation, having begun when the Utility formed in 1997.
What is the background or condition for the need for the practice?	Seattle Public Utilities has 1200 employees and a $500 million operating budget. Their ten-year capital program exceeds $1 billion. The organization consists of: ▪ Water Supply ▪ Drainage ▪ Wastewater ▪ Solid Waste ▪ Engineering Services As a result of the 1997 reorganization, the Seattle Public Utilities was created to provide integrated utility services to Seattle citizens. The creation of the new utility provided opportunities to rebuild, improve and integrate existing business and human systems. Visionary leadership created a formula for success in the Utility, and a Mayor who is committed to measurable results. This formula provided a commitment of resources and money to build the Utilities' Organizational Performance Framework. The Organizational Performance Framework consists of four components: 5. Strategic Business Planning: Setting and defining vision, mission, values, goals, and results in a 2-year strategic business plan.

Utility:	**Seattle Public Utilities**
Person Contacted:	**Joanne Peterson, Dove Albert, Cyndy Holtz**
Practice:	**Alignment of an Organizational Performance System**
	6. Operational Planning: Develop work unit performance measures and targets, operational work plans, and a budget. 7. Performance Measurement: Collect feedback (customer, financial organizational, operational), measure and evaluate. Products include a performance measurement database, customer feedback system, and a reporting system. 8. Performance Management: Use data to manage change, make decisions, reallocate resources. Create/adapt business systems and continually develop workforce skills, knowledge, and leadership. Products include utility-wide leadership core competencies, 360-degree management evaluations, employee surveys, leadership and employee performance and development planning, management performance contracts and employee performance reviews linked to performance measures.
How was it implemented, and who was involved in the implementation?	**STRATEGIC PLANNING** The first strategic planning process conducted in early 1997 was a top down, bottom up approach and produced a 2-year (1997-1998) Strategic Business Plan. Over 250 employees from department head to line worker were involved in creating the Utilities' vision, mission, goals and objectives. Employees identified 50 strategic efforts that were linked to accomplishing the objectives and were published in the Plan. At the end of 1998, the Strategic Effort Result Report identified over 95% of the efforts had met the targeted completion dates. The second strategic planning process was another top down and bottom up approach. The 9-member executive team set the strategic direction for the department, reviewing the 1997-1998 Strategic Goals, developing result statements and aligning the Utilities' Strategic Business Plan with the Mayor's "Managing for Results" program—G.A.R.M. (Goals, Actions, Results and Measures). The bottom up approach was focused on involving employees at the work unit level in the development of performance measures and connecting their everyday work with the strategic goals of the Utility. To involve employees, a Strategic Business Plan campaign with a performance-focused message was kicked off at a March, 1999 Utility Leadership Forum with over 150 Utility management staff – from department director to line crew supervisor. The Mayor attended as guest speaker to talk about how the work of the Utility supported the goals of the City and what the Managing for Results program was all about. The Utilities Director reinforced the important messages about performance and results, using the Utilities 1999-2000 Strategic Business Plan to make the connection. A series of 5 supervisor workshops was conducted in March, 1999 to further educate line supervisors about the Strategic Business Plan and prepare them to serve as key communicators of the Strategic Business Plan to their employees and help them develop meaningful

Utility:	**Seattle Public Utilities**
Person Contacted:	**Joanne Peterson, Dove Albert, Cyndy Holtz**
Practice:	**Alignment of an Organizational Performance System**

performance measures for their unit's work. The workshops provided the supervisors with tools to better understand each of the six strategic goals and how they could help employees connect their everyday work to the Utilities strategic goals. The supervisors received baseball hats in recognition of their important role of communicating the Plan to employees.

During the month of April, supervisors conducted staff meetings with their employees to roll out the Strategic Business Plan. Every employee was asked to demonstrate how his or her work connected to a strategic goal. Each employee received a copy of the Plan and a Strategic Goals Note pad.

A monthly news article in the SPU News features an employee interview about how his or her work connects to a strategic goal of the Utility. An internal web site for the Utilities' Strategic Business Plan is available to employees and will link to the operational work unit performance measures at a future date.

OPERATIONAL PLANNING

In February, 1998 we began the operational planning process. The existing Utilities 1998 Work Program served as the starting point to develop the G.A.R.M. report. The goals, actions, and results were revised through the Strategic Business Planning process.

PERFORMANCE MEASUREMENT

The process to develop performance measures began in summer 1998. An operational work plan format was developed in alignment with the required G.A.R.M. report.

A two-day Utilities workshop with the 39-member leadership team, senior budget and policy staff (approximately 30) and appointed performance deputies (6) was held in October, 1998 to conduct performance measurement training, and begin development of unit performance measures using budget activities.

Over the next 3 months, work units developed 1999-2000 operational work plans that included performance measures and initiatives. Work units set targets, identified baselines (where possible), established data sources and assigned data managers for each of the performance measures. Executive Team reviews occurred at regular intervals with the final operational work plans approved in January 1999.

Next year we will focus on acquiring customer feedback and looking into the benchmarking and best practices of other utilities.

PERFORMANCE MANAGEMENT

Performance management includes defining the Utilities' culture, developing leadership to manage change, and recognizing, rewarding

Utility:	Seattle Public Utilities
Person Contacted:	**Joanne Peterson, Dove Albert, Cyndy Holtz**
Practice:	**Alignment of an Organizational Performance System**
	and developing employees. It is the "human side" of performance. Several projects have served to set the stage to move forward in the performance management process. In Jan 1999, the Utilities Executive Team established a set of leadership core competencies and conducted a pilot 360-degree evaluation with its 39-member leadership team. In July 1999, an employee opinion survey, the "Employee Voice" was conducted to set a baseline of employee satisfaction. The items were based on the Organizational Values and Work Principles. In 1997, performance contracts for senior management and key policy staff were developed that aligned with the Strategic Business Plan. In 1998, the contract system was refined to include performance measures. In 1999, the employee performance review system was revised to align with the Strategic Business Plan, include the leadership core competencies for supervisors and evaluate the knowledge and usage of performance measures.
Who are the owners of the practice (or the process)?	The owners of the Utilities' Organizational Performance practice are the Utilities Executive Team.
What systems or tools support the practice?	The performance measures data base went into design in May 1999. Data input began in August and is targeted for completion in October. Training of department staff on the database will occur in the last quarter of 1999. It is expected the data managers or a designee will derive the performance data for the 12-month actuals from the database.
What other departments or systems are effected by the practice, and what are the interactions?	Performance sessions were conducted with the Department Director during July, 1999. Six-month actuals were presented, targets and performance measures were discussed and refined over the next several weeks. Performance sessions, using the operational work plans, will occur in bi-annually with the Department Director to discuss performance, targets, needed adjustments/refinements. The Utilities G.A.R.M. was reported in July 1999 to the City Council and Mayor. The G.A.R.M. will be reported bi-annually. Additional reporting will occur at 12, 18 and 24-month intervals. In 2000, an SPU customer scorecard will be designed and distributed to interested stakeholders.
What are the results/benefits/outcomes of the practice?	▪ An integrated, connected approach to continuous improvement. ▪ A system that recognizes performance, motivates employees and inspires improvement. ▪ Employees who are focused on the work that is most important.
How do you actually use the results/benefits/ outcomes of	▪ 2001-2002 Budget decisions will be made with input from the performance as reported in the operational work plans.

Utility:	Seattle Public Utilities
Person Contacted:	Joanne Peterson, Dove Albert, Cyndy Holtz
Practice:	Alignment of an Organizational Performance System
the practice?	▪ Work units identify areas needing improvement and undertake work process redesign efforts to increase efficiency. ▪ Reallocate resources and budget.
How do you measure success of the practice, and is there a formal instrument for measuring?	The formal instruments are the Operational Work Plan and the GARM report. There will also be a 2000 Citizen Scorecard, which will be mailed to all Seattle citizens and other stakeholders.
Were any of the results surprising, and if so, what were they?	Lessons learned: ▪ It's all about people and their willingness to take ownership and use the system ▪ It needs to add value ▪ It needs to be constantly communicated ▪ It's a continuous change process ▪ Build on what you already have ▪ It doesn't have to be perfect the first time around; be flexible ▪ Watch what you measure because you may get an unexpected result (i.e. number of calls answered within ___ seconds. Revenues may fall because call reps are giving customers a quick yes to get them off the phone.)
What barriers do you see that might affect future success of the practice?	▪ A new mayor or Department Director with a different focus ▪ A majority of new council members, and a new Utilities and Environmental Management Committee chair who has different ideas, etc. ▪ Internal resources get reallocated away from project ▪ System is perceived to be cumbersome and is not supported by the line supervisors, managers and executive management ▪ Consistent, constant communication throughout the organization helping the employees connect their work to the strategic goals does not happen

Utility: Louisville Water Company	
Person Contacted: Beverly Irons, Director of Workforce Diversity (formerly Corporate Quality Coordinator)	
Practice: Participation – Working in Teams	
Describe the practice.	Louisville Water Company has established and maintained a practice of working teams throughout the Utility, moving from a traditional hierarchical based organization to one where self-directed teams set and achieve common goals. This includes long-term work teams, as well as shorter-term project related teams.
How long has it been in place, and what stage is it in (i.e., assess, vision, design, implement)?	Teams have been in place since 1992, when first efforts were made to move to working in teams. The team concept is now well established and fully implemented throughout the Company.
What is the background or condition for the need for the practice?	In 1992, the Company made the decision to become a team-based organization, as a result of new leadership in the Company and an overall organizational change effort. In 1993, they recognized that doing this required significant effort in establishing expectations and goals, and providing the appropriate tools and training.
How was it implemented, and who was involved in the implementation?	An outside consultant was hired to provide initial training to three cross-functional project teams, starting with the operational areas. The Company President, John Huber, provided input and involvement, and "drove" the process. The consultant helped the teams learn about roles and responsibilities, how to conduct and lead meetings, and provided a learning process for managers to let go and encourage teams to work toward becoming self-directed.
Who are the owners of the practice (or the process)?	John Huber, the President, and the Executive Leadership Team own the quality initiative. Natural Work Teams (NWTs) are part of the LWC systems-based management structure. Business system owners and Process Owners along with the NWTs practice systems-based management and are accountable for the delivery of products and services.
What systems or tools support the practice?	Measures have been established and are continuing to be developed. Over the past three years, everyone in the organization has received team training. The Human Resources Department assumed the responsibility for training from the consultant, and currently trains all new hires.
What other departments or systems are effected by the practice, and what are the interactions?	The entire Company is involved, and working as teams has become embedded in the culture of the organization. As part of the organization change effort that was begun in 1992, 15 Business Systems were formed, each having a System Advisory Council. These teams provide input to managers.
What are the	▪ Higher quality work life

Utility:	Louisville Water Company
Person Contacted:	**Beverly Irons, Director of Workforce Diversity (formerly Corporate Quality Coordinator)**
Practice:	**Participation – Working in Teams**
results/benefits/outcomes of the practice?	▪ Employees feel that they are making meaningful contributions to the Company, rather than feeling helpless and devalued as they have in the past ▪ People have grown professionally and are more involved in their work ▪ There is open communication and shared decision-making
How do you actually use the results/benefits/ outcomes of the practice?	Results are used for planning efforts. They are also used by teams to assess how they are working together, in addition to assessing their work processes for improvements and accomplishment of goals.
How do you measure success of the practice, and is there a formal instrument for measuring?	*LWC has a corporate scoreboard or a "corporate family of measures" that was developed internally, and is still evolving, based on key results areas (KRAs).* The number of employees involved in teams is tracked, and the process improvements that teams work on, i.e., what was produced, what worked, what happened to the results, who chartered the team, who removed the barriers so that the team could be more effective. Work is still being done on measuring dollar savings, and plans are to develop this in 2000. Measurements will be integrated in order to make decisions based on the data analysis.
Were any of the results surprising, and if so, what were they?	"The results were not especially surprising – we expected to be successful. But, it was gratifying to really see that it happened." There was resistance from some – especially managers – it was a learning process for them to let go in order for teams to really become self-directed. Time and training helped managers to support the practice.
What barriers do you see that might affect future success of the practice?	"We don't see any real barriers – the organization culture has changed and even if the project driver were no longer there, we would not revert."
Additional information.	Another program, the "Facilitators Services Program," started as a result of the teams process. There are 22 trained facilitators who spend 15 percent of their time facilitating teams.

Utility:	**Louisville Water Company (LWC)**
Person Contacted:	**Mike Sturgeon**
Practice:	**Compensation – Performance Recognition**
Describe the practice.	LWC has restructured and modified percentages for performance recognition. Their Work Planning and Performance Management system (WPPM) has put the market driven structure in place. The program has grades of 1-14, with market value and range as a control point for each grade.
How long has it been in place, and what stage is it in (i.e., assess, vision, design, implement)?	The practice was established in April 1999.
What is the background or condition for the need for the practice?	Since 1991-92, employees have established their own goals for the year. The employees are evaluated against those goals and given a rating for their wage increases. There was criticism from employees that there were not enough increases and goals were not standard.
How was it implemented, and who was involved in the implementation?	The Human Resource Business System is the owner and was responsible for implementation of the practice.
Who are the owners of the practice (or the process)?	
What systems or tools support the practice?	Data reflective of market movement and WPPM results.
What other departments or systems are effected by the practice, and what are the interactions?	All systems are eligible and all employees can participate.
What are the results/benefits/outcomes of the practice?	People feel better about their work – they are motivated by the market-driven structure and can accelerate toward market.
How do you actually use the results/benefits/ outcomes of the practice?	It has increased motivation and helped to retain employees.
How do you measure success of the practice, and is there a formal instrument for measuring?	The WPPM has developed a quartile system to measure performance and establish a rating for the grades.
Were any of the results surprising, and if so, what were they?	The program is in infant stages with no data as yet.
What barriers do you see that might affect future success of the practice?	Proper utilization of the WPPM system.

Utility:	**Louisville Water Company (LWC)**	
Person Contacted:	**Mike Sturgeon**	
Practice:	**Compensation – Performance Recognition**	
Additional information.		In addition to the market-driven performance recognition model, there are three new ways to reward employees. These include: Token System: Each year, every employee is given 12 tokens with a value of $2 each. The tokens are to be given on the spot to fellow employees for recognition of exceptional service or notable achievements. The tokens can be accumulated by recipients and redeemed at local restaurants, discount stores, etc. Charter System: Each year, each employee can be recognized if they participated in a charter team, i.e., a baseball team, and someone recommends them for recognition. There is a formal chartering process with criteria such as having a sponsor and a facilitator. Vouchers for $25 are given to each member of the team, up to $250 per team. Quality Celebration System: Once a year employees can be recognized up to $12 per employee to celebrate what they were able to do to promote quality. Employees must write up what they want to do, i.e., some form of recreation like mini-golf, etc. or a team building exercise, and submit the form to the Quality Office.

Utility:	Louisville Water Company (LWC)
Person Contacted:	**Mike Sturgeon**
Practice:	**Union Relations**
Describe the practice.	LWC uses a systems-based approach to labor relations. The utility has built a partnership with the union.
How long has it been in place, and what stage is it in (i.e., assess, vision, design, implement)?	The practice was started in 1993-94, and is in the implementation stage. The program has received several local awards, including one from the Labor Management Center of the University of Louisville.
What is the background or condition for the need for the practice?	In the late 1980s and early 1990s, the utility and the union had an adversarial relationship. This was evidenced by a high number of grievances and arbitrations. The union couldn't get the work they wanted, and the utility couldn't get the cooperation or efficiency that it wanted. A systems-based management approach was instigated at LWC. This moved them from utility "departments," to a water utility based on systems, processes, and quality. The cornerstone of this approach is a partnership of all employees, which vastly improved the disconnect between the union and the company. They realized that they must form a partnership in order for everyone to be successful.
How was it implemented, and who was involved in the implementation?	A Decision Partner Team (DPT) comprised of union and company representatives was put in place. Union representatives include the President and Vice President of the union and six other chairs. Recognizing the need to have front line workers on board, the six union representatives are selected by the union executive board. The company has fewer chairs, and includes the Business System Owner of Human Relations (consisting of the processes of labor and employee relations, compensation and benefits, recruitment, development and retention of employees. It is important to note that the DPT does not cover grievances or arbitration issues, but looks at opportunities within the organization to further develop the partnership between the company and the union.
Who are the owners of the practice (or the process)?	The project Sponsor is the Core Business System Owner, plus the HR Business System Owner, and the Union President and Vice President.
What systems or tools support the practice?	The Systems Based Management Approach, adopted by LWC and used throughout the utility. The Decision Partner Team (DPT), comprised of union and company representatives. The Union Quality Coordinator position.
What other departments or systems are effected by the practice, and what are the	Everyone in the utility is affected by the practice. The DPT meets weekly. They do not get into labor-related issues, but work together to enhance communications efforts.

Utility:	Louisville Water Company (LWC)
Person Contacted:	**Mike Sturgeon**
Practice:	**Union Relations**
interactions?	
What are the results/benefits/outcomes of the practice?	Significant reduction in grievances, arbitrations and union issues. Also, membership in the union has increased by 20 percent. It was recognized that the union lacked consistent direction due to going through an election process every two years. Union members held a retreat to develop their own Vision and Mission as well as a Strategic Plan to support the company's Strategic Plan.
How do you actually use the results/benefits/ outcomes of the practice?	One example is that a "Union Quality Coordinator" position was created. This is a person selected by the union who has a chair on the DPT and spends 100% of their time looking for opportunities to improve quality.
How do you measure success of the practice, and is there a formal instrument for measuring?	The DPT identified current levels of involvement, and where they wanted to be, and developed a plan for how to get there. Using a scale of 1 – 5, level of involvement is measured quarterly and annually against the original baseline.
Were any of the results surprising, and if so, what were they?	The union people felt they wanted greater participation for their own comfort level. Once they got involved through the company's invitation, they found that they really didn't need as much involvement as they thought. Likewise, the company people felt rather threatened or undermined that union people had to be involved. Both sides matured and came to an understanding for the appropriate level of involvement.
What barriers do you see that might affect future success of the practice?	The unknown; years into this process there may be union elements that won't buy into this practice. Continuing change in demographics; new people continually need to be educated. The aggressiveness of the union; the timelines for a practice like this is 10-15 years, but the union wants success in as little as six months.
Additional information.	

Utility:	**City of Portland Water Bureau**
Person Contacted:	**Susan Bailey, Program Manager for Workforce Development**
Practice:	**Communications - Knowledge Transfer**
Describe the practice.	This practice involves transferring knowledge from a large number of potential retirees to the newer workforce so that institutionalized personal knowledge isn't lost when people retire.
How long has it been in place, and what stage is it in (i.e., assess, vision, design, implement)?	The practice is in a design phase and has grown out of the Water Bureau's larger workforce planning project.
What is the background or condition for the need for the practice?	The Water Bureau wanted to have a systematic approach to workforce planning. The utility has a large number of people nearing retirement and needed to plan for filling vacancies. They also wanted to redefine positions in order to create a more flexible and diverse workforce with the level of skills needed to do the work. As a result of the workforce planning project it was recognized that a more formalized process needed to be in place in order to capture and retain information that might leave as individuals retire from the utility.
How was it implemented, and who was involved in the implementation?	The practice is currently in a Design phase - each work unit is working toward capturing personal information, experience, and knowledge so that it becomes part of the organization. *New employees are being hired before the person they are replacing leaves, so that training and knowledge transfer can occur as new people learn skills from their predecessors and work into their positions. Cross-training of existing staff is also occurring. Procedural manuals, standardized record keeping formats, archiving and data retrieval systems are also be designed and implemented* Retirees are being rehired on a part time basis to train new employees.
Who are the owners of the practice (or the process)?	There is Bureau-wide ownership of this practice, as it is evident that all work units face similar numbers of potential retiring employees.
What systems or tools support the practice?	Videos are being made of people doing their work, and work units are documenting personal information, experience, and knowledge so that it becomes part of the organization. A new GIS (Geographic Information System) is coming on line soon that will help to support this practice. Also, a new CIS (Customer Information System) is being implemented that will allow the utility to capture additional information about specific meters and addresses.
What other departments or systems are effected by the practice, and what are the interactions?	There is Bureau-wide involvement in this practice as all departments have a large number of potential retirees.
What are the results/benefits/outcomes of the	Better training is happening. There is an increased awareness of the need to capture knowledge and experience so it is available to transfer

Utility:	**City of Portland Water Bureau**
Person Contacted:	**Susan Bailey, Program Manager for Workforce Development**
Practice:	**Communications - Knowledge Transfer**
practice?	to others. Better designed documentation and record keeping systems.
How do you actually use the results/benefits/ outcomes of the practice?	The primary benefit is that institutionalized knowledge isn't lost as people leave or retire.
How do you measure success of the practice, and is there a formal instrument for measuring?	There is no formal instrument for measuring at this time.
Were any of the results surprising, and if so, what were they?	
What barriers do you see that might affect future success of the practice?	
Additional information.	

Utility:	**City of Portland Water Bureau**
Person Contacted:	**Susan Bailey, Program Manager for Workforce Development**
Practice:	**Utility Worker Classification – Water Service Utility Worker Apprentice Program**
Describe the practice.	This program is designed to allow for the hiring of unskilled employees and to bring them to journey level within two years through a highly structured training program. *This practice creates flexible utility workers via a formal 4000-hour State-approved apprentice program using training and mentoring. This is an official apprenticeship program overseen by the City of Portland's Trades Apprenticeship Joint Labor and Management Committee. The program provides skilled and flexible utility workers, and also enhances the City's labor and management partnering.*
How long has it been in place, and what stage is it in (i.e., assess, vision, design, implement)?	The process has taken approximately 18 to 24 months from the beginning concept of the program and task analysis, to hiring the first six apprentices as of September 9, 1999. It is now in the first stages of implementation.
What is the background or condition for the need for the practice?	The Water Bureau wanted to have a systematic approach to workforce planning. The utility has a large number of people nearing retirement and needed to plan for filling vacancies. They also wanted to redefine positions in order to create a more flexible and diverse workforce with the level of skills needed to do the work.
How was it implemented, and who was involved in the implementation?	After the beginning concept of flexible workers, a thorough job and task analysis of utility worker positions was done, with the help of a consultant hired through a local community college. The process of seeking and getting approval by the State of Oregon and City of Portland took a long time because of organizational changes and the fact that this would be a new position in the City. Also, the City itself was in the middle of a large job re-classification study. *Susan Bailey, the Water Bureau's Program Manager for Workforce Development, was a member of the group doing the City's Apprenticeship Taskforce (AFT), a group tasked with creating apprentice opportunities within the city. Mort Anoushiravani, a Senior Manager at the Bureau, was instrumental in fostering the Water Bureau's flexible worker position. Susan worked with the ATF, State of Oregon Bureau of Labor and Industries, the City's Trades Apprenticeship Committee, and community based organizations, many of which represent minorities and women who want to get into the trades.*
Who are the owners of the practice (or the process)?	The Water Bureau. Also, through the process of developing the program, it has become a "show-case" project for the City.
What systems or tools support the practice?	The system that supports the practice is the formal State-approved apprentice program. Record keeping is done using various spreadsheet applications that are required because it is a State program.

Utility:	**City of Portland Water Bureau**
Person Contacted:	**Susan Bailey, Program Manager for Workforce Development**
Practice:	**Utility Worker Classification – Water Service Utility Worker Apprentice Program**
What other departments or systems are effected by the practice, and what are the interactions?	The City is using the program as a "show case" project that has increased the diversity of City employees. Currently the City's Bureau of Human Resources is working with the Water Bureau and the ATF to enhance recruitment and selection practices using the approach developed for this program as a model.
What are the results/benefits/outcomes of the practice?	▪ *A systematic approach to workforce planning.* ▪ A more diverse pool of applicants. ▪ The potential to have trained and flexible workers with the necessary skills to fill needed positions as they occur. ▪ Enhanced recruitment and selection procedures that utilize community resources to create a more diversified applicant pool. ▪ Enhanced position for the City as an employer of choice.
How do you actually use the results/benefits/ outcomes of the practice?	Retiring workers are being re-hired on a part-time basis to serve as trainers and mentors to apprentices. The first six apprentices were hired as of September 9, 1999.
How do you measure success of the practice, and is there a formal instrument for measuring?	Measurement is indistinct at this time. As people complete the program, numbers can be bench-marked and compared, and the diversity of the workforce can be measured. A better trained workforce is harder to measure. Statistically, the first six apprentices hired show greater diversity than previous hirings in similar job classes. The applicant pool as a whole is more diversified than pools for similar job classes.
Were any of the results surprising, and if so, what were they?	As the program developed it was discovered that people who complete the apprenticeship process could serve as a "feeder" positions for other areas as skills develop and mature.
What barriers do you see that might affect future success of the practice?	A program like this needs to be started early on to prepare for anticipated employee retirement.
Additional information.	Because this program will be brining in large numbers of new employees, many of whom have less experience, are younger and more diverse, the Bureau is developing programs to help ensure their successful integration into the workforce. Existing journey level employees and supervisors have been trained in on-the-job training and coaching techniques. Training in creating a respectful workplace is being designed for all employees, and a mentoring program for all new or transitioning employees is being developed.

Utility:	Regional Municipality of Ottawa-Carleton, Water Division
Person Contacted:	**André Proulx**
Practice:	**Work Process Redesign – Water Distribution Excavation**
Describe the practice.	Water distribution system excavation process, which includes all aspects of planned and emergency maintenance when an excavation is required to perform the work.
How long has it been in place, and what stage is it in (i.e., assess, vision, design, implement)?	The process took 6 months to complete. The practice has been implemented (about 80 %) since Dec. 1998. Further cross training of staff is still required.
What is the background or condition for the need for the practice?	Based on metric benchmarking (similar to Xerox method) the utility's excavation process was not considered best in class. An 8-member team was formed to study the process and improve it.
How was it implemented, and who was involved in the implementation?	The process involved the 8-member team, and took 6 months to complete. The team developed detailed action plans that described actions, responsible persons, monitoring and costs. They went on 2 site visits to see some best in class processes in operation, and held debriefing sessions while still on the road.
Who are the owners of the practice (or the process)?	The 8 team members own the process. The project was championed by high-level management (Director) and this involvement is seen as critical to the success of the project.
What systems or tools support the practice?	High-level management support is critical, along with providing the dollars needed to perform site visits. Recommendations from this process resulted in transfer of staff, cross training and adopting new equipment, i.e., a backhoe with extended reach. Following this practice, management staffed a full time training coordinator. As well, a training room with 3 computers was set up on site for computer based training.
What other departments or systems are effected by the practice, and what are the interactions?	Our Corporate Call Center and Corporate Fleet personnel got involved. Recommendations included transferring the Corporate fleet operators to the water division to become "true" team members of the excavation crew, and to become cross trained.
What are the results/benefits/outcomes of the practice?	*Better work planning and reduction of equipment were a direct benefit. Better staff morale, and the ability for staff to do more kinds of work allowed for the reduction in crew size, while providing management with more flexible staff. Staff's involvement in the process led to better accountability and ownership of the process.* Direct cost savings: $210,000/yr Increased productivity: $540,000/yr These savings take into account the Training Coordinator, and the computer training room, but do not take into account the purchase of new backhoes. These will be purchased as the older ones are replaced.
How do you actually use the	Management makes sure ideas get implemented; they are included in

Utility:	Regional Municipality of Ottawa-Carleton, Water Division
Person Contacted:	**André Proulx**
Practice:	**Work Process Redesign – Water Distribution Excavation**
results/benefits/ outcomes of the practice?	long-term planning, and short-term recommendations are made for training and necessary equipment purchases. A mid-management champion is assigned the responsibility to make certain the practices/recommendations are being implemented over a specific time period.
How do you measure success of the practice, and is there a formal instrument for measuring?	▪ Reduction in staff numbers (moved from a 7 to a 5 person crew). ▪ Reduction in material use. ▪ Time on site (it's hard to measure this without long-term results). ▪ Actual reduction in budgets and work order cost. Activity based costing (ABC) is required to monitor this appropriately. The region does not have a good ABC system to date and looks to purchase one in the year 2000 or 2001. Currently, our water distribution computerized maintenance management system captures most of these costs.
Were any of the results surprising, and if so, what were they?	The overall process identified that frequently, the length of down time at the site was surprising. This down time was attributed to poor planning, and included improper equipment arriving at site, wrong materials, locates (for other utilities) not yet completed, to name a few.
What barriers do you see that might affect future success of the practice?	A reduction of staff is imminent when best practices show a reduction of 2 staff per crew should take place. People fear losing jobs. Due to the age of our workforce, we were able to meet this objective through attrition (people retire or leave). Cross-training to use people to the best of their abilities is needed, but there is some objection from some staff. Pilot training programs with crews willing to try something new has worked well for the Region.
Additional information.	Lessons learned: ▪ Teams must be representative of the functional areas ▪ Focus on action ▪ Team must have ownership of the process ▪ Teams must meet deadlines ▪ The process and implementation must have a "champion" ▪ Selection of benchmarking partners is critical ▪ Training essential to develop "in-house" expertise ▪ Management must be committed to implement recommendations ▪ Financial support for the process must be in place

APPENDIX C
SURVEY RESULTS—BAR CHARTS

Category: Respondent Profile
QA - Tenure/Longevity - Years with Utility

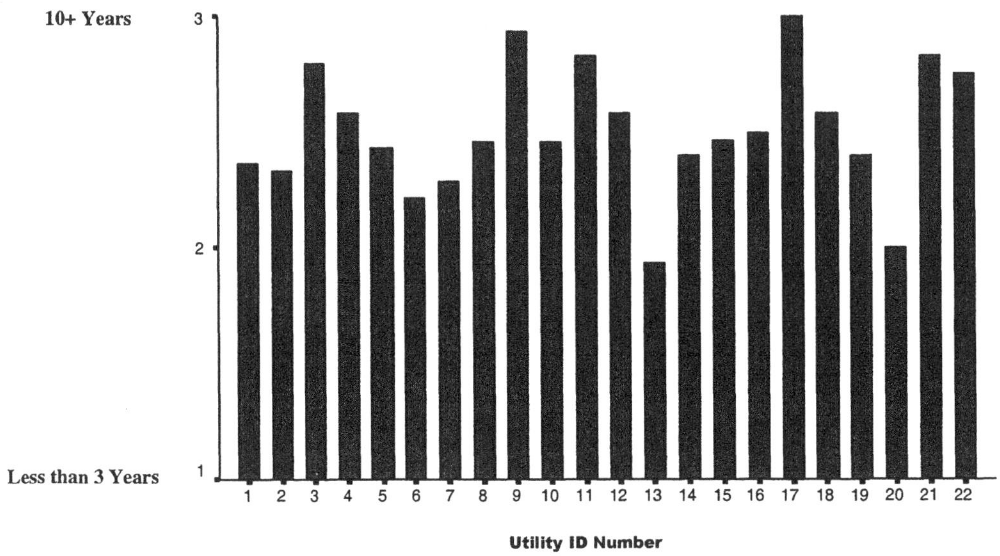

Overall Statistics			
Mean	2.5	Minimum	1
Median	3.0	Maximum	3
Std. Dev.	.6	Population	22

Category: Respondent Profile
QB - Labor Representation

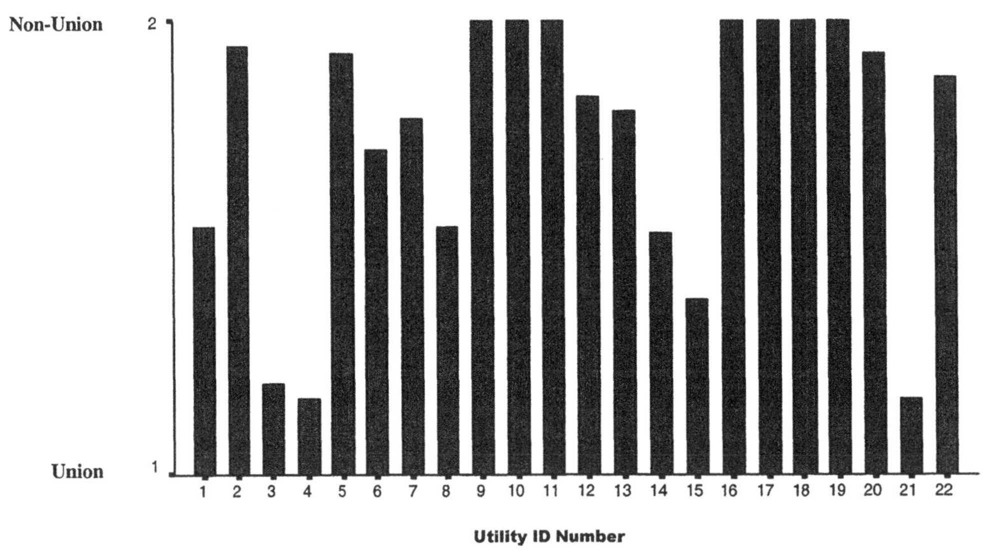

Overall Statistics			
Mean	1.7	Minimum	1
Median	2.0	Maximum	2
Std. Dev.	.4	Population	22

Category: Respondent Profile
QC - Level within Organization

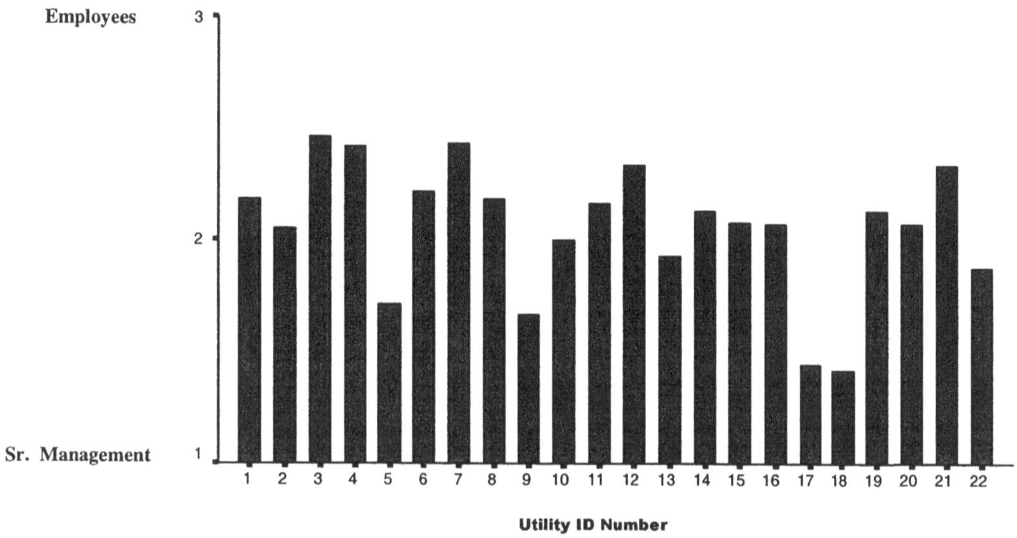

Overall Statistics			
Mean	2.1	Minimum	1
Median	2.0	Maximum	3
Std. Dev.	.7	Population	22

Category: Respondent Profile
QD - Job Function in the Utility

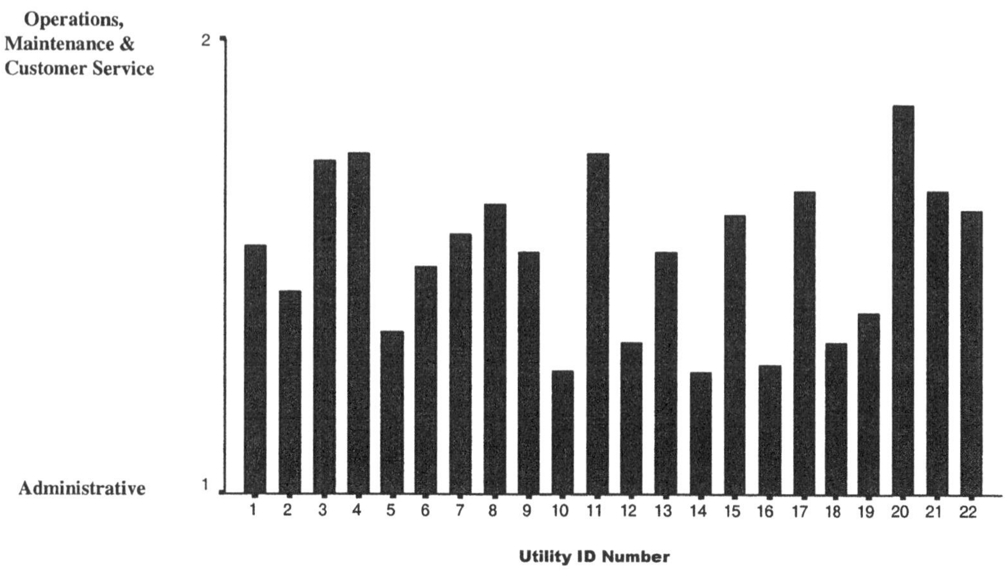

Overall Statistics			
Mean	1.5	Minimum	1
Median	2.0	Maximum	2
Std. Dev.	.5	Population	22

Category: Alignment
Q1 - This utility has established and communicated its mission, vision, and values

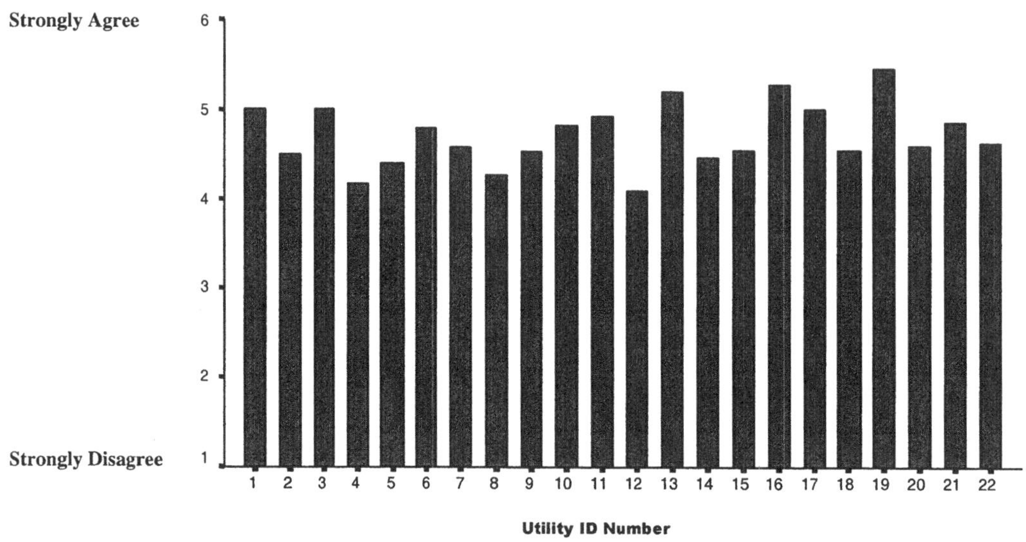

Overall Statistics			
Mean	4.7	Minimum	1
Median	5.0	Maximum	6
Std. Dev.	.9	Population	22

Category: Alignment
Q2 - I see evidence that established mission, vision, and values are effective in guiding this utility

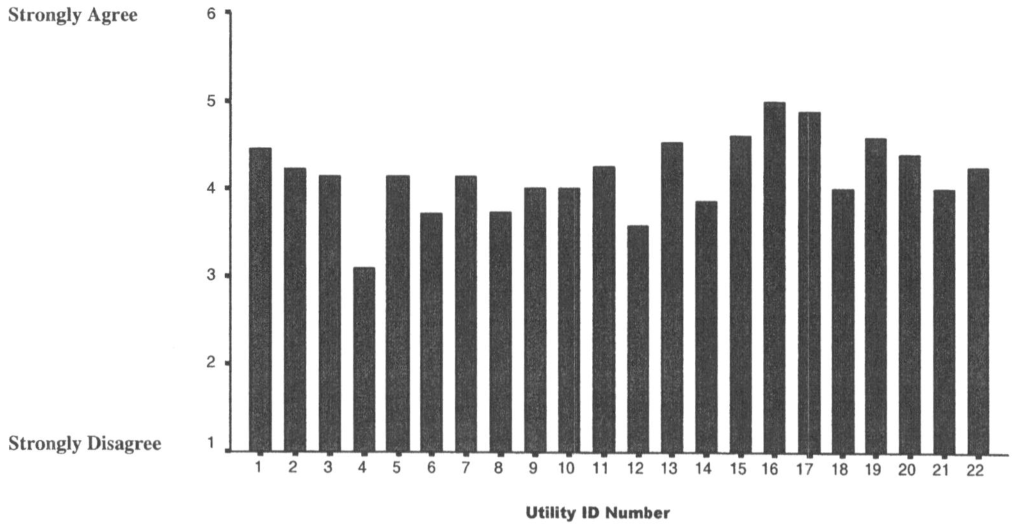

Overall Statistics			
Mean	4.2	Minimum	1
Median	4.0	Maximum	6
Std. Dev.	1.0	Population	22

Category: Alignment
Q3 - Adequate resources are committed to achieve long-term goals

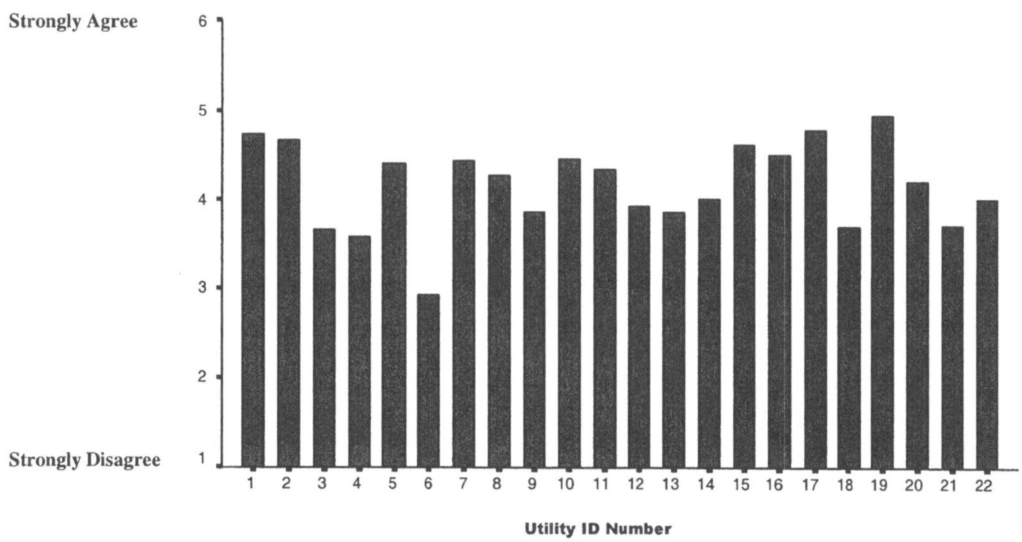

Overall Statistics			
Mean	4.2	Minimum	1
Median	4.0	Maximum	6
Std. Dev.	1.1	Population	22

Category: Alignment
Q4 - I believe that my job should include "whatever it takes" to satisfy our customers

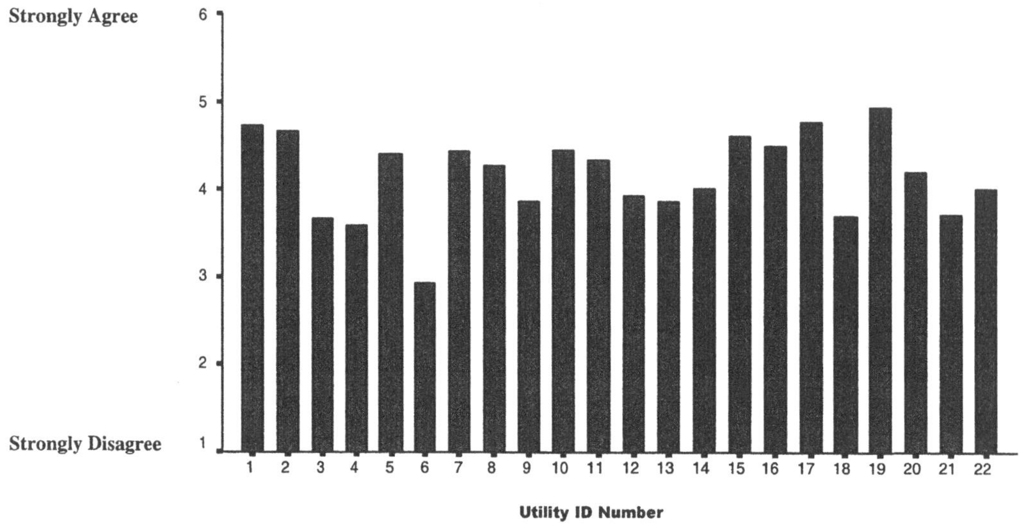

Overall Statistics			
Mean	4.8	Minimum	1
Median	5.0	Maximum	6
Std. Dev.	1.0	Population	22

Category: Alignment
Q41 - This utility emphasizes the creation and communication of long-term goals

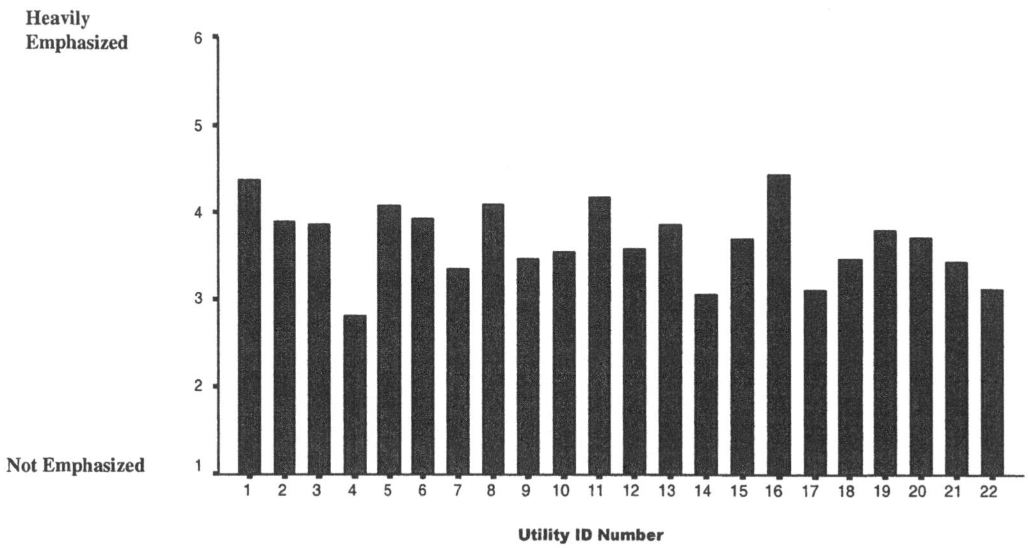

Overall Statistics			
Mean	3.7	Minimum	1
Median	4.0	Maximum	6
Std. Dev.	1.1	Population	22

Category: Alignment
Q42 - This utility emphasizes that all employees deliver a high level of customer service

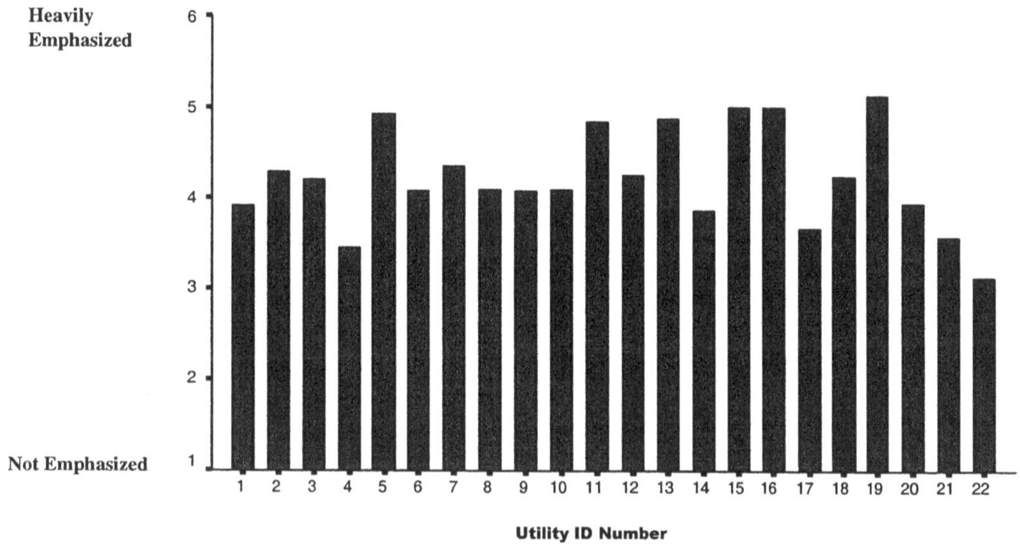

Overall Statistics

Mean	4.2	Minimum	1
Median	4.0	Maximum	6
Std. Dev.	1.2	Population	22

Category: Alignment
Q46 - This utility has achieved alignment between labor/union and management/utility interests

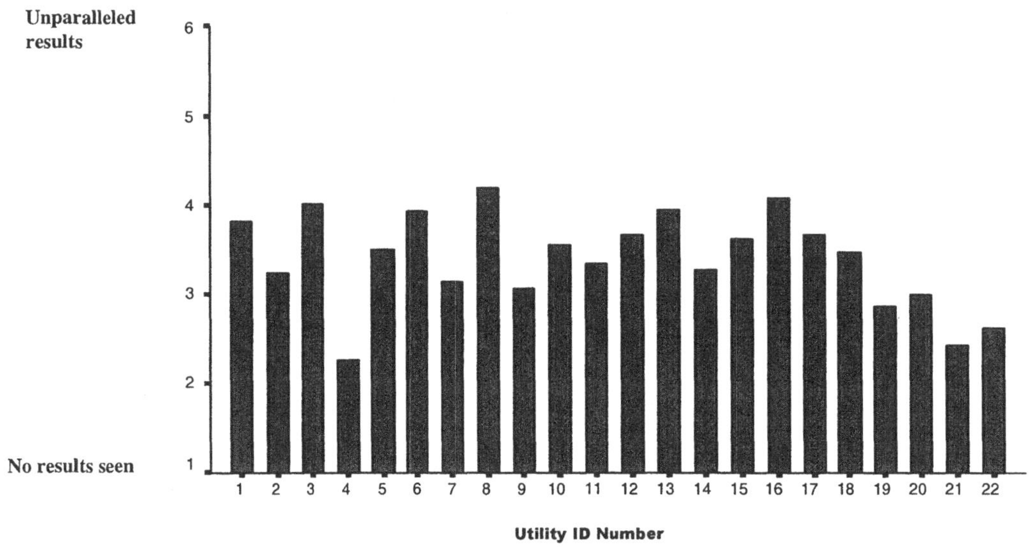

Overall Statistics			
Mean	3.4	Minimum	1
Median	4.0	Maximum	6
Std. Dev.	1.1	Population	22

Category: Alignment
Q79 - I believe it is important that there is alignment of my goals with organizational long-term goals

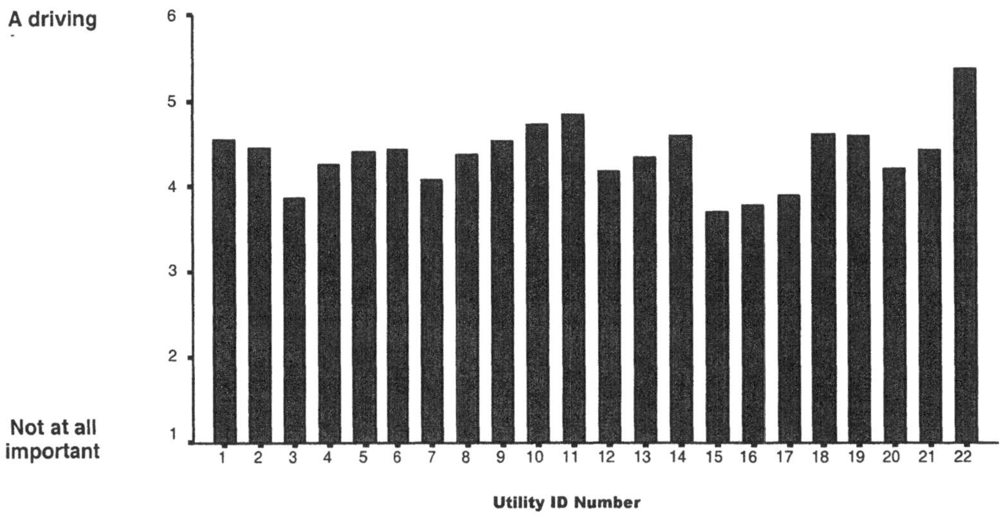

Overall			
Mean	4.4	Minimum	1
Median	5.0	Maximum	6
Std. Dev.	1.1	Population	22

Category: Alignment
Q80 - It is important to have common problem/solving processes/methods at this utility

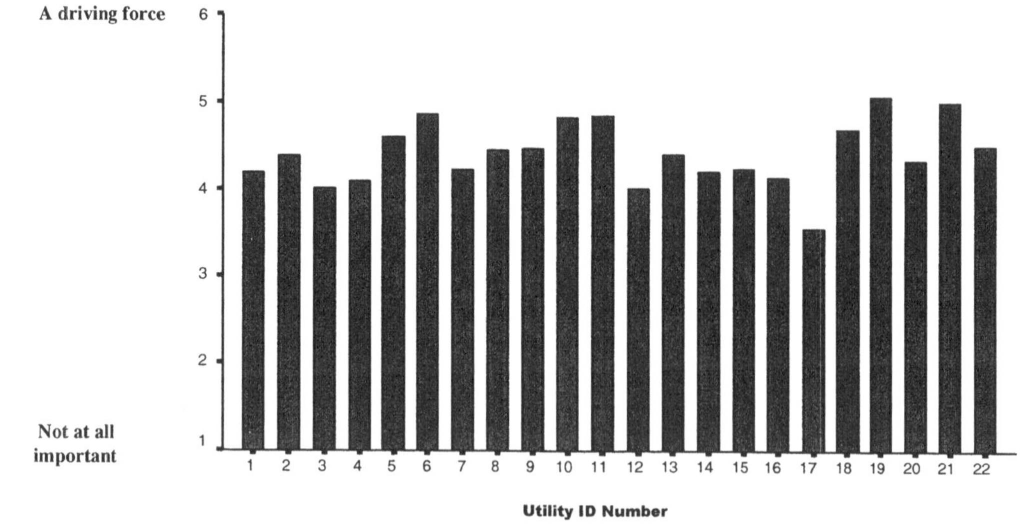

Overall Statistics			
Mean	4.4	Minimum	1
Median	5.0	Maximum	6
Std. Dev.	1.0	Population	22

Category: Urgency
Q5 - Competitive market/industry forces are significant drivers of planned change initiatives

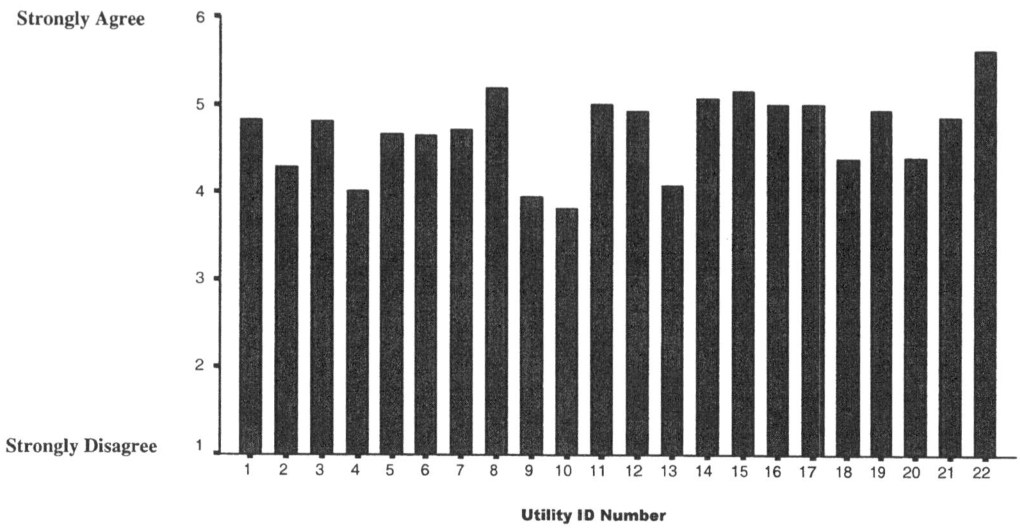

Overall Statistics			
Mean	4.7	Minimum	1
Median	5.0	Maximum	6
Std. Dev.	1.1	Population	22

Category: Urgency
Q6 - Stakeholders are significant drivers of planned change initiatives

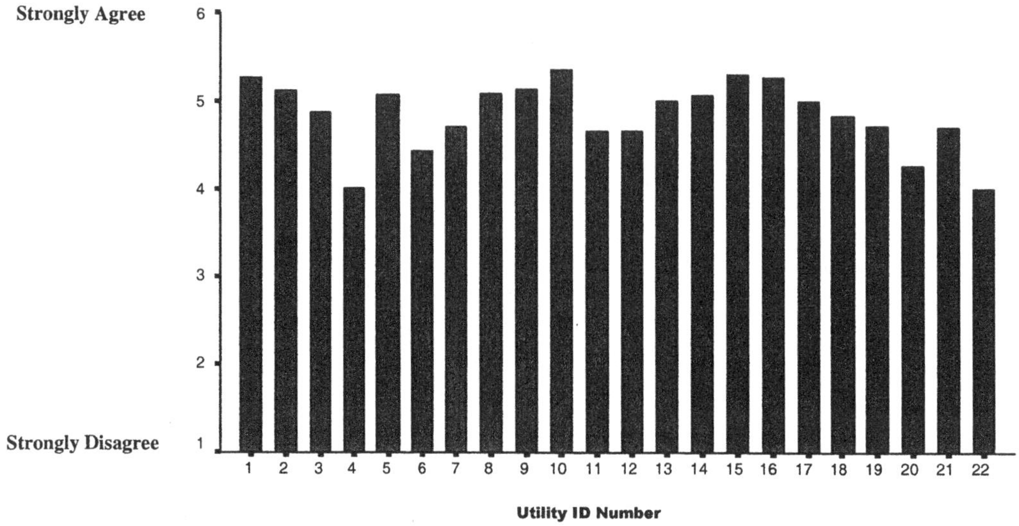

Overall Statistics			
Mean	4.8	Minimum	1
Median	5.0	Maximum	6
Std. Dev.	1.0	Population	22

Category: Urgency
Q7 - This utility recognizes the need to redesign work practices to improve business performance and customer satisfaction

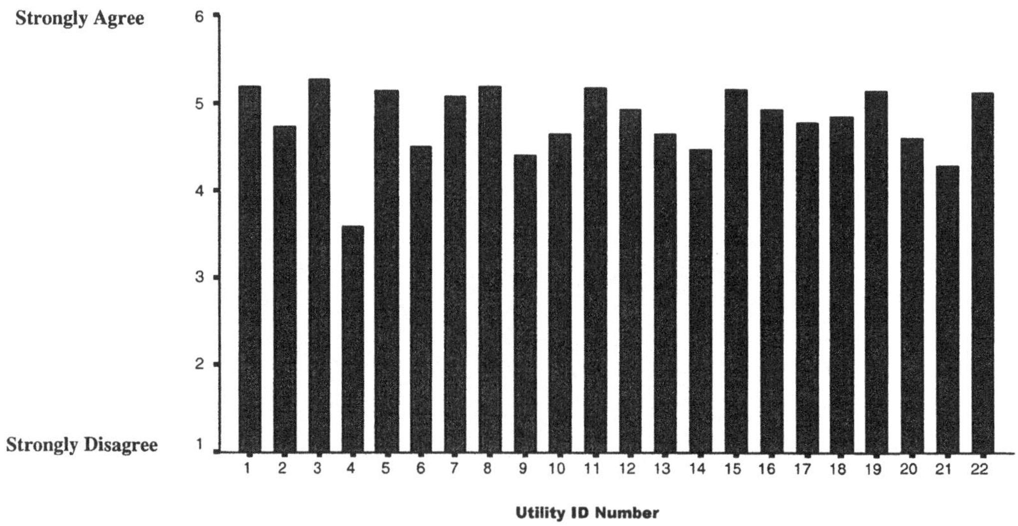

Overall Statistics			
Mean	4.8	Minimum	1
Median	5.0	Maximum	6
Std. Dev.	1.0	Population	22

Category: Urgency
Q8 - This utility is currently engaged in planning initiatives for major change

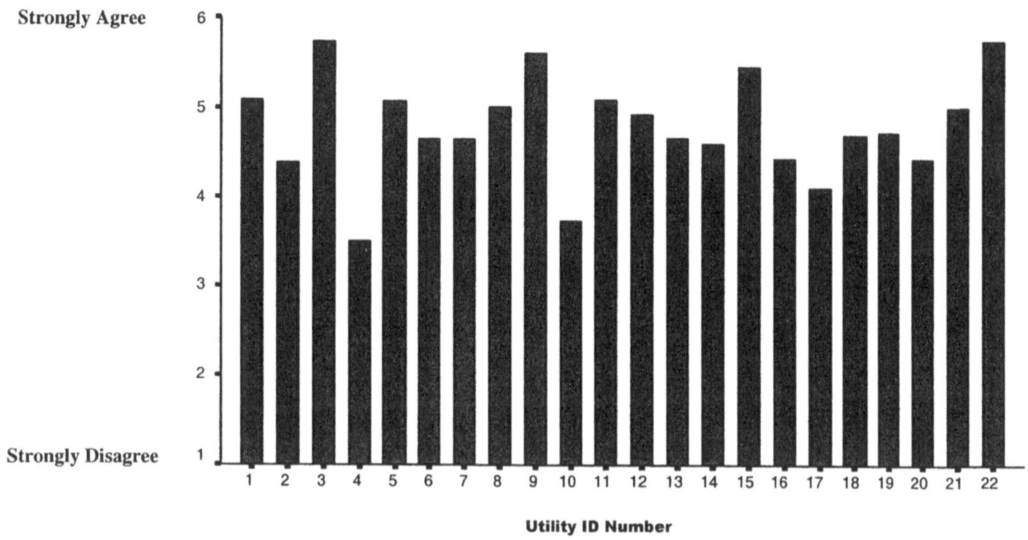

Overall Statistics			
Mean	4.8	Minimum	1
Median	5.0	Maximum	6
Std. Dev.	1.1	Population	22

Category: Urgency
Q74 - It is my perception that ___% of this utility's customers are dissatisfied with the services they receive

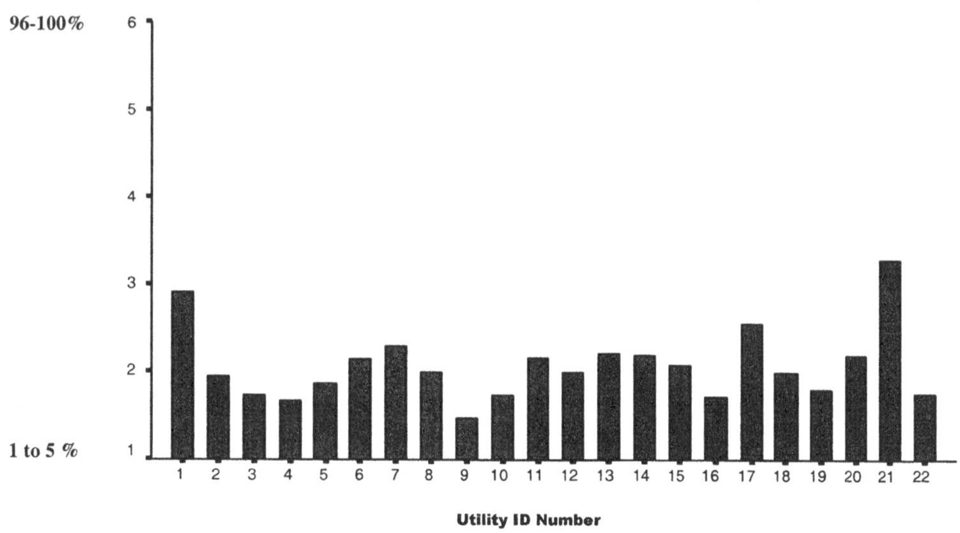

Overall Statistics			
Mean	2.1	Minimum	1
Median	2.0	Maximum	6
Std. Dev.	1.2	Population	22

Category: Urgency

Q75 - It is my perception that ___% of this utility's employees are dissatisfied with the way the utility is run

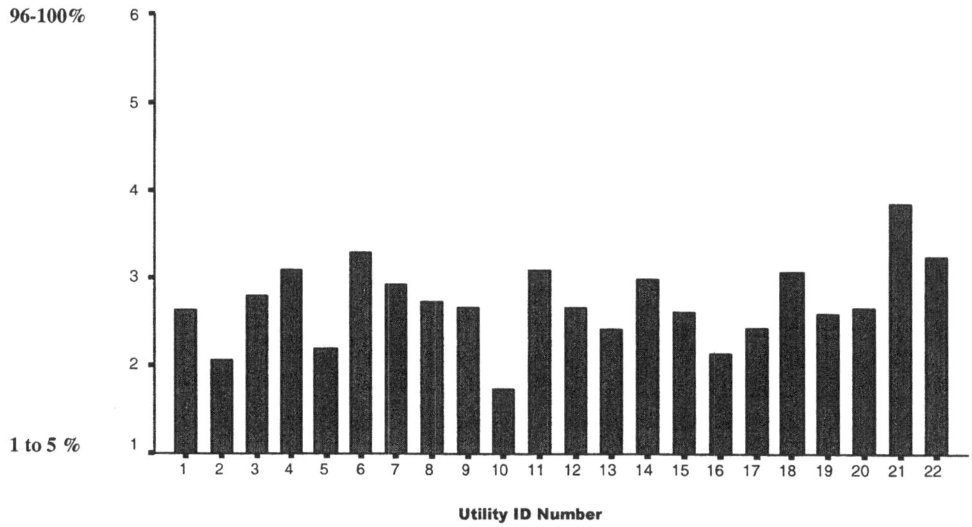

Overall Statistics			
Mean	2.7	Minimum	1
Median	2.8	Maximum	6
Std. Dev.	1.2	Population	22

Category: Urgency

Q81 - It is important for this utility to have a plan in place to deal with competitive pressures

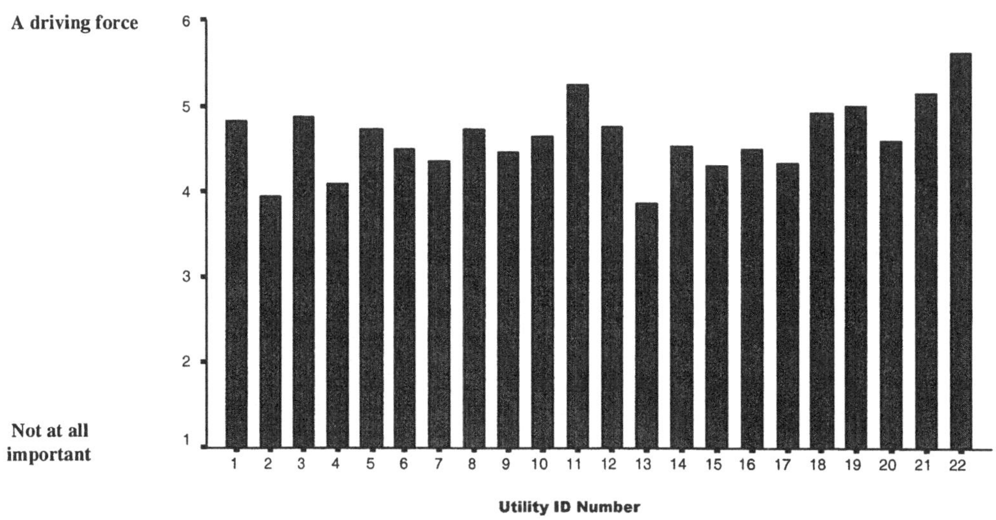

Overall Statistics			
Mean	4.6	Minimum	1
Median	5.0	Maximum	6
Std. Dev.	1.1	Population	22

Category: Leadership
Q9 - Leadership of this utility is visibly committed to improvement

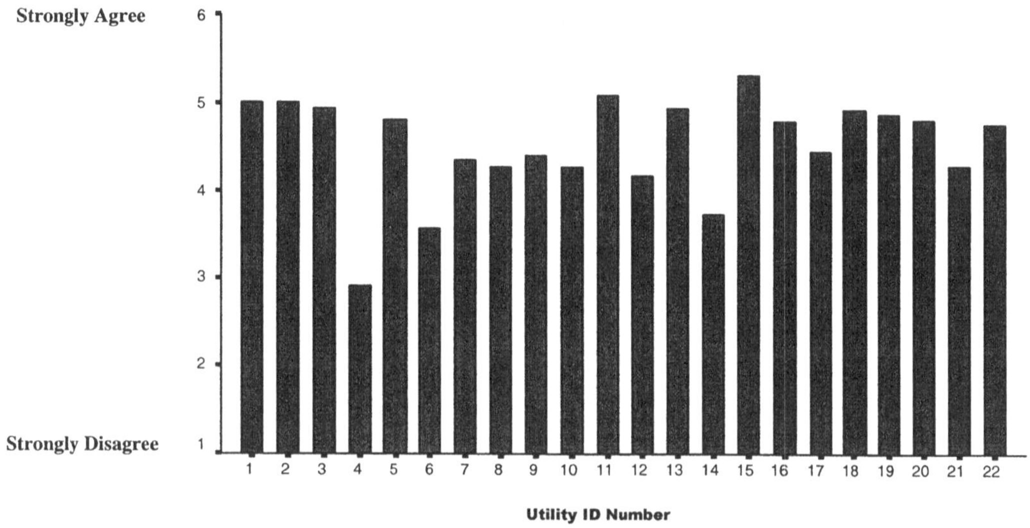

Overall Statistics			
Mean	4.5	Minimum	1
Median	5.0	Maximum	6
Std. Dev.	1.2	Population	22

Category: Leadership
Q10 - Leadership of this utility does what they say they will do, or "walks their talk"

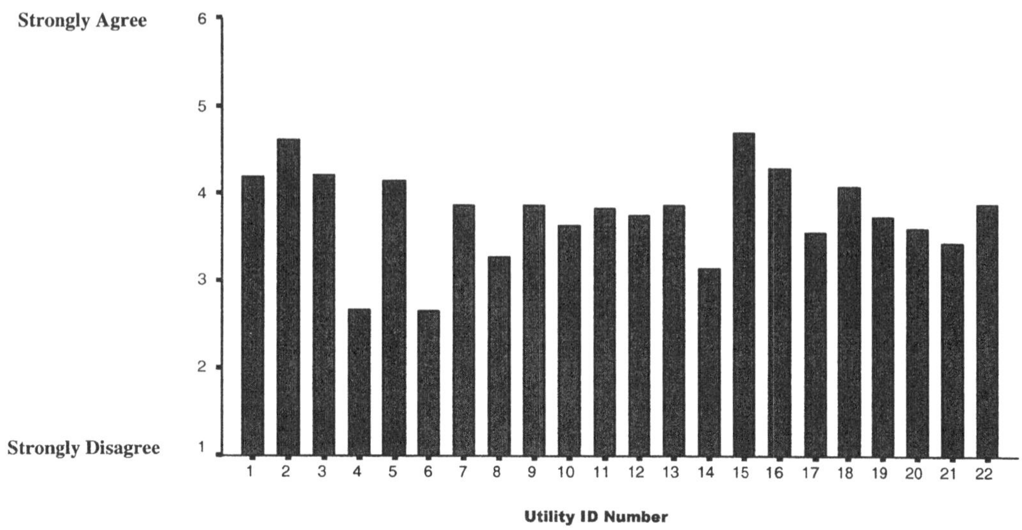

Overall Statistics			
Mean	3.8	Minimum	1
Median	4.0	Maximum	6
Std. Dev.	1.2	Population	22

Category: Leadership
Q43 - This utility emphasizes the development of leadership

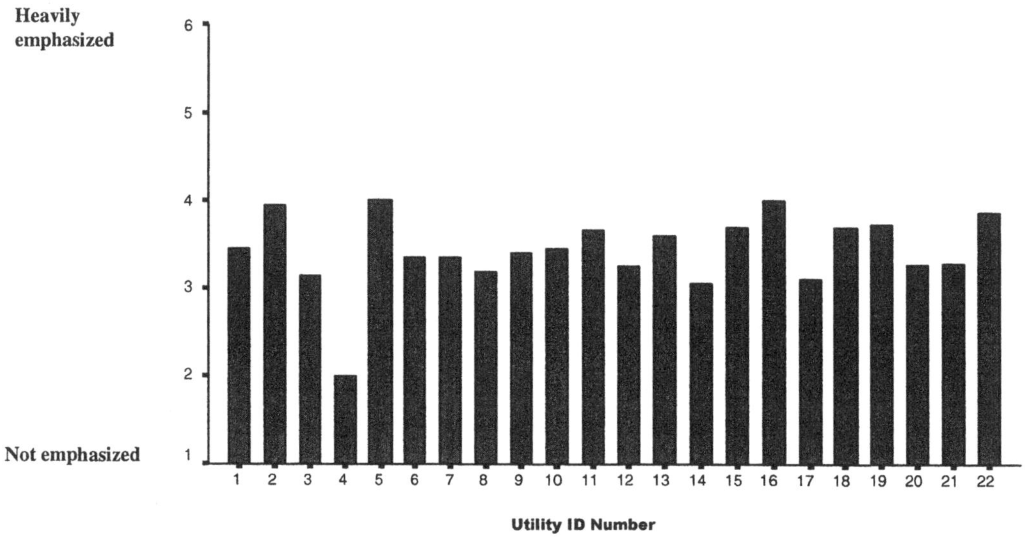

Overall Statistics			
Mean	3.4	Minimum	1
Median	4.0	Maximum	6
Std. Dev.	1.2	Population	22

Category: Leadership
Q51 - The practice of coaching or mentoring is included as an important part of every employee's job responsibilities

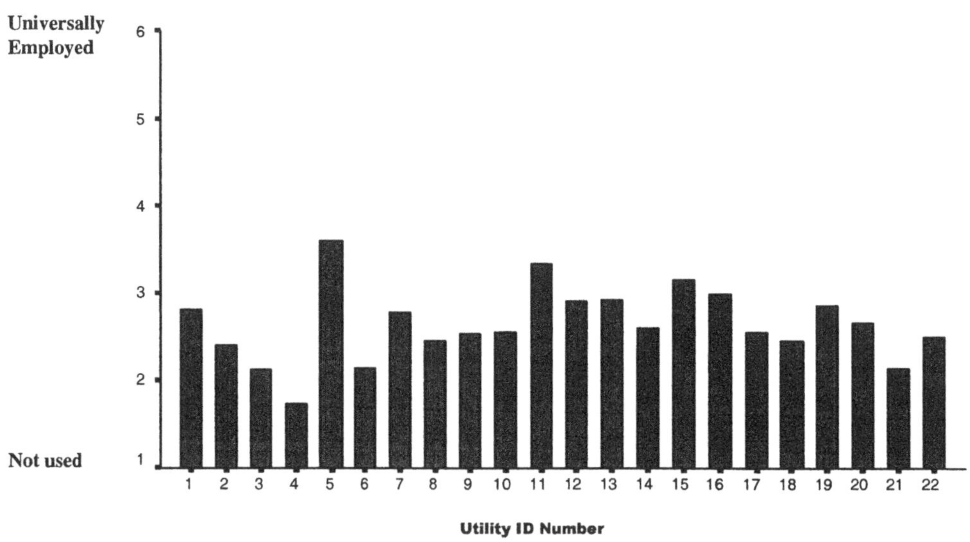

Overall Statistics			
Mean	2.7	Minimum	1
Median	3.0	Maximum	6
Std. Dev.	1.3	Population	22

Category: Leadership
Q62 - In planning initiatives for change, the leadership of this utility considers the resulting effect on employees

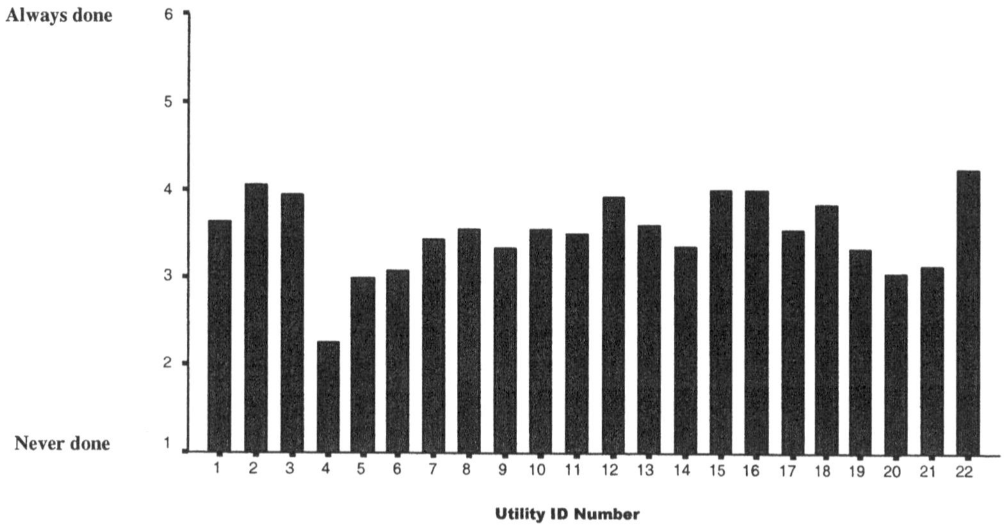

Overall Statistics			
Mean	3.5	Minimum	1
Median	3.0	Maximum	6
Std. Dev.	1.3	Population	22

Category: Decision-Making
Q11 - It is clear at what level decisions are made at this utility

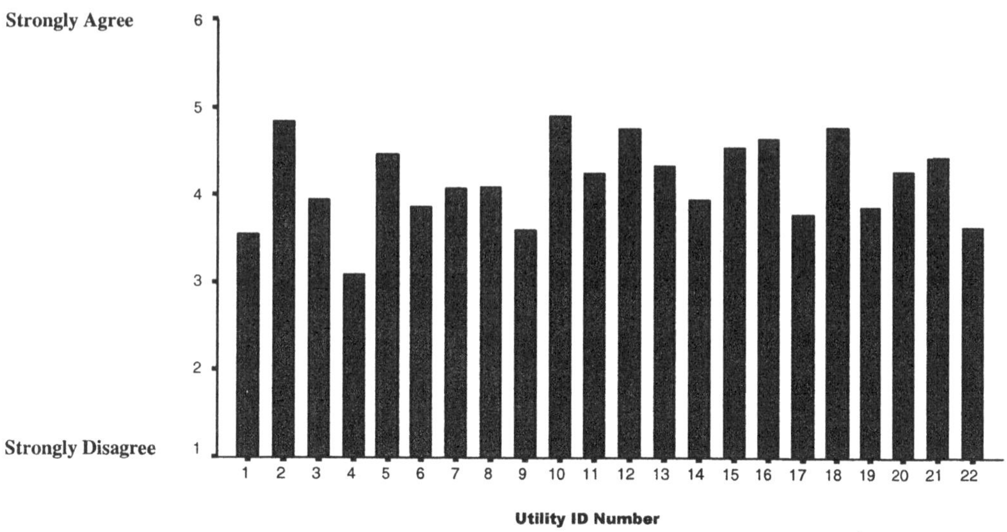

Overall Statistics			
Mean	4.2	Minimum	1
Median	4.0	Maximum	6
Std. Dev.	1.1	Population	22

Category: Decision-Making
Q12 - Individuals at this utility readily accept accountability and responsibility for decision-making

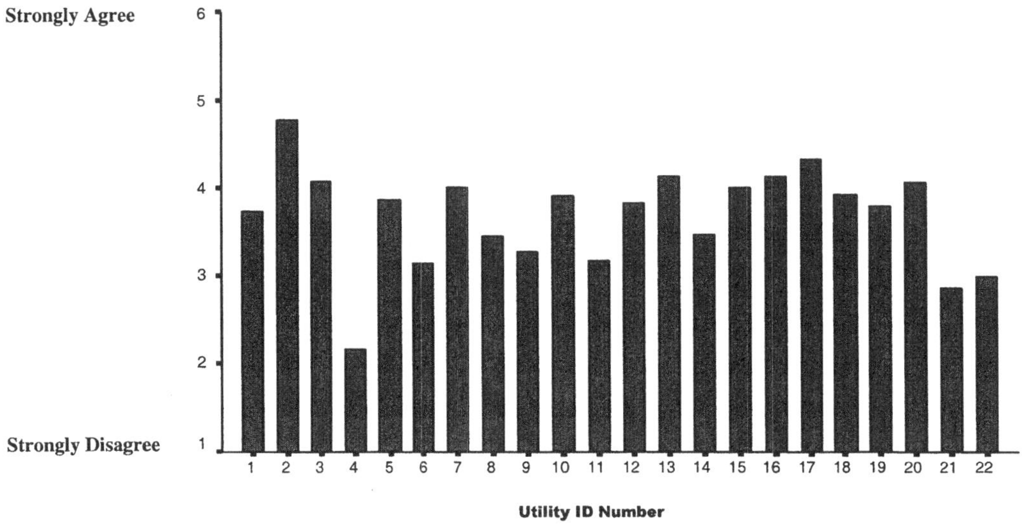

Overall Statistics			
Mean	3.7	Minimum	1
Median	4.0	Maximum	6
Std. Dev.	1.2	Population	22

Category: Decision-Making
Q47 - This utility has been successful in delegating most day-to-day decisions beyond management/supervisors

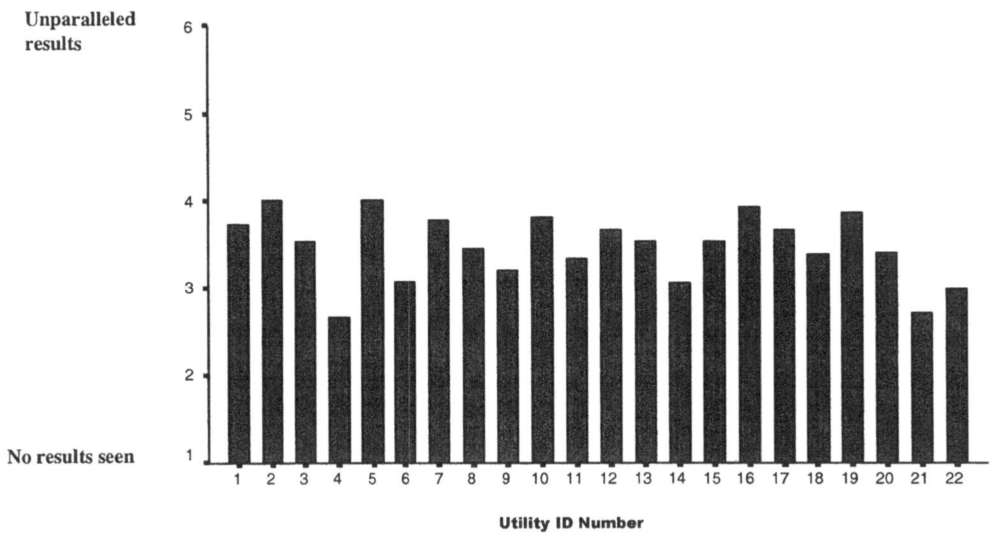

Overall Statistics			
Mean	3.5	Minimum	1
Median	4.0	Maximum	6
Std. Dev.	1.0	Population	22

Category: Decision-Making
Q82 - It is important at this utility that deregulation of authority for day-to-day decisions is at the lowest possible level

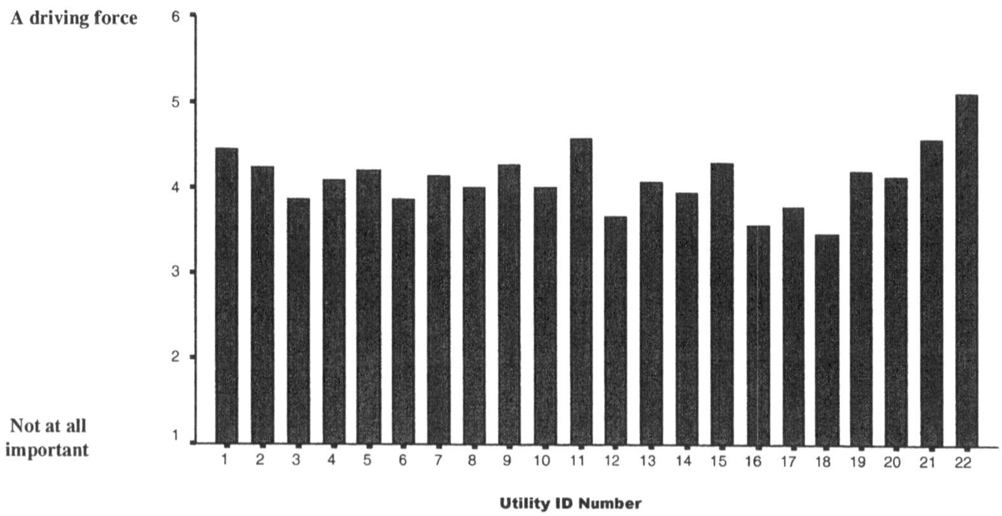

Overall Statistics			
Mean	4.1	Minimum	1
Median	4.0	Maximum	6
Std. Dev.	1.2	Population	22

Category: Participation/Style
Q13 - My input is actively sought and applied in improving this utility's performance

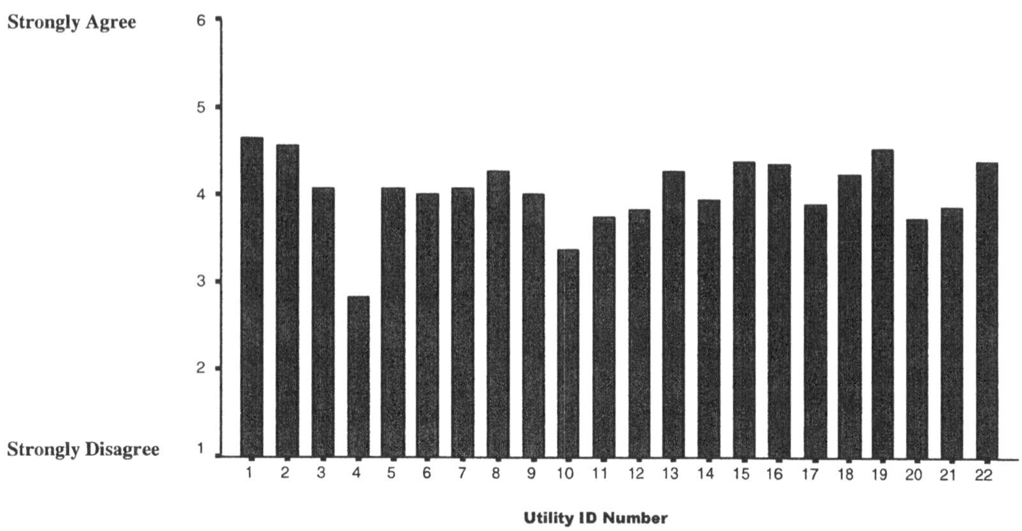

Overall Statistics			
Mean	4.1	Minimum	1
Median	4.0	Maximum	6
Std. Dev.	1.2	Population	22

Category: Participation/Style
Q14 - Roles and responsibilities are clear at this utility

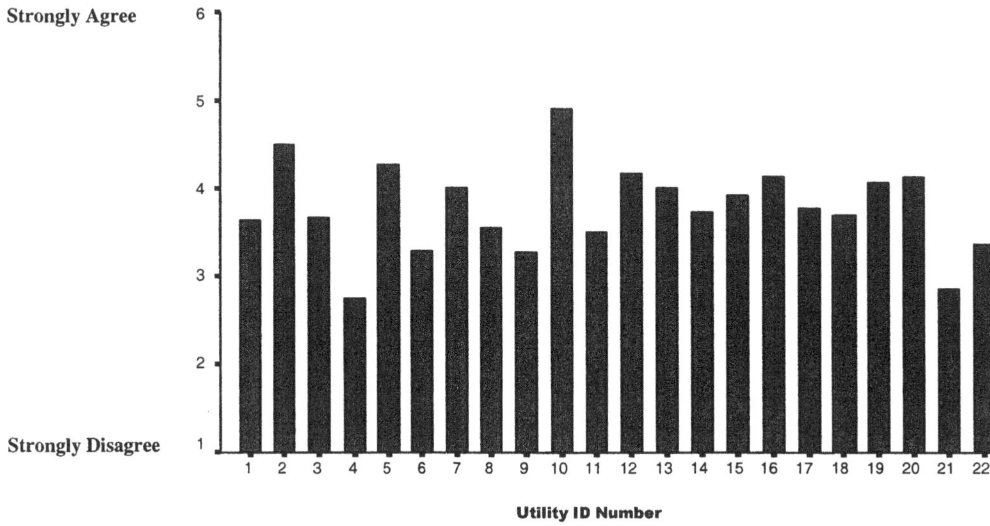

Overall Statistics			
Mean	3.8	Minimum	1
Median	4.0	Maximum	6
Std. Dev.	1.1	Population	22

Category: Participation/Style
Q15 - Employees view themselves as essential members of a team

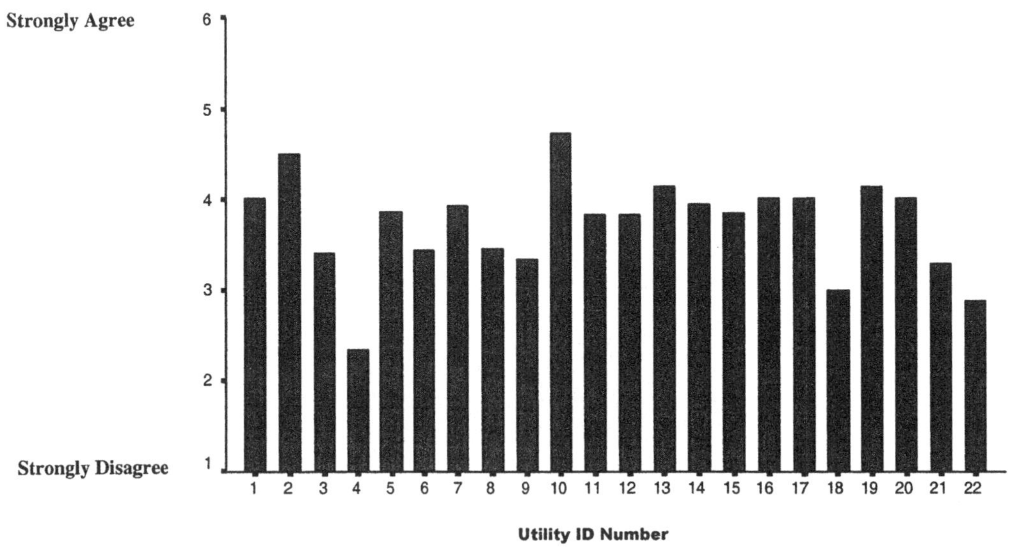

Overall Statistics			
Mean	3.7	Minimum	1
Median	4.0	Maximum	6
Std. Dev.	1.1	Population	22

Category: Participation/Style
Q16 - Labor and management at this utility work together to set up new job classifications and pay systems

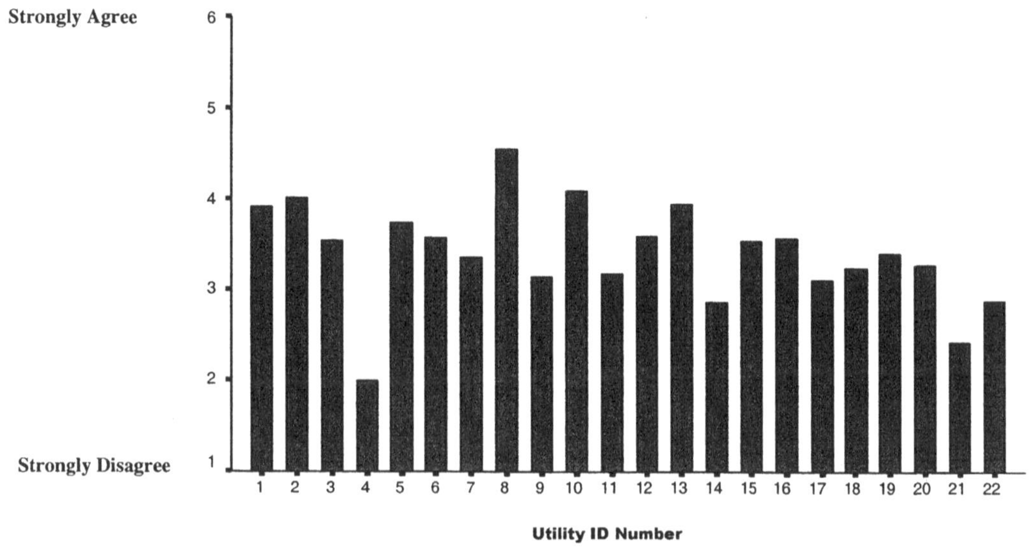

Overall Statistics			
Mean	3.4	Minimum	1
Median	4.0	Maximum	6
Std. Dev.	1.2	Population	22

Category: Participation/Style
Q17 - Labor and management at this utility collaboratively establish incentives for customer and business improvement goals

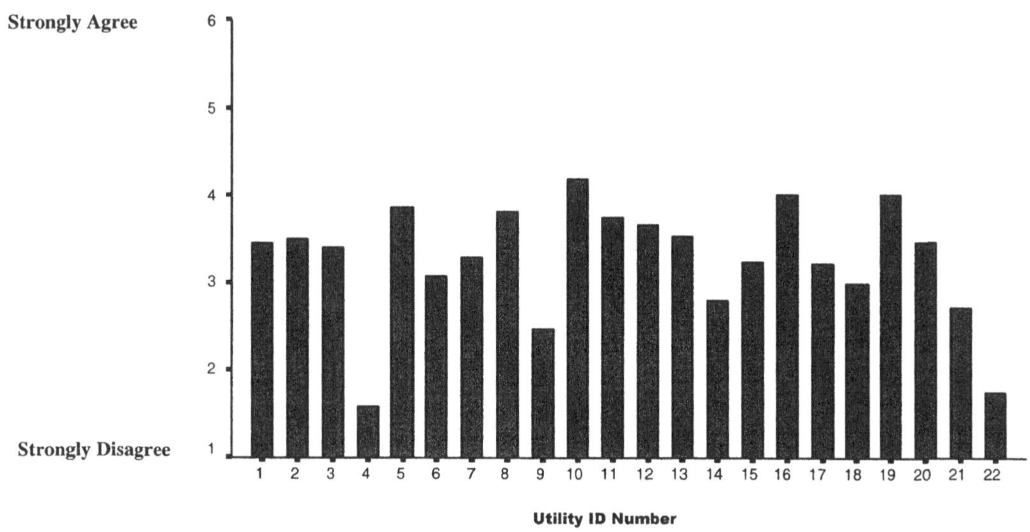

Overall Statistics			
Mean	3.3	Minimum	1
Median	3.0	Maximum	6
Std. Dev.	1.3	Population	22

Category: Participation/Style
Q44 - All employees at this utility are encouraged to take personal initiative

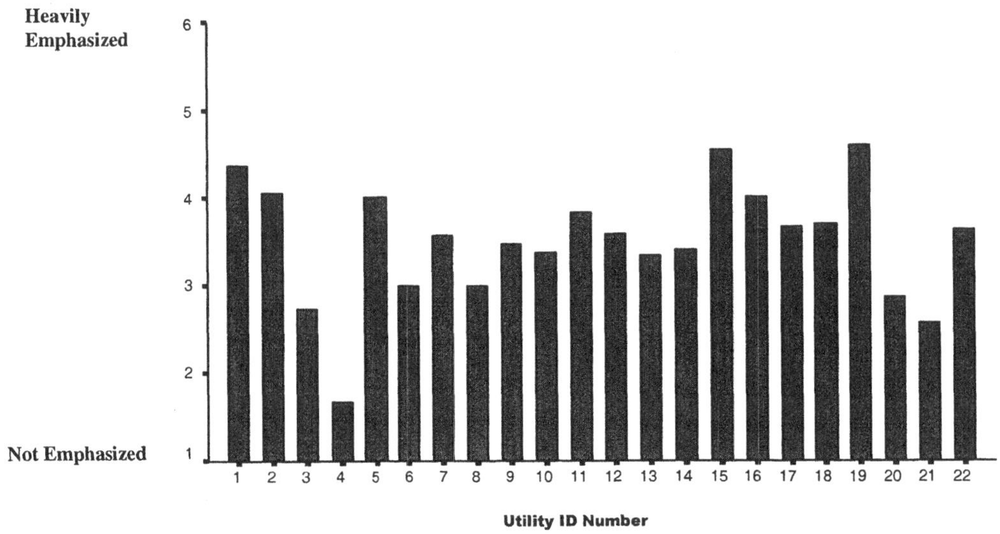

Overall Statistics			
Mean	3.5	Minimum	1
Median	4.0	Maximum	6
Std. Dev.	1.4	Population	22

Category: Participation/Style
Q48 - Use of team-based approaches in this utility has resulted in improved business processes and work practices

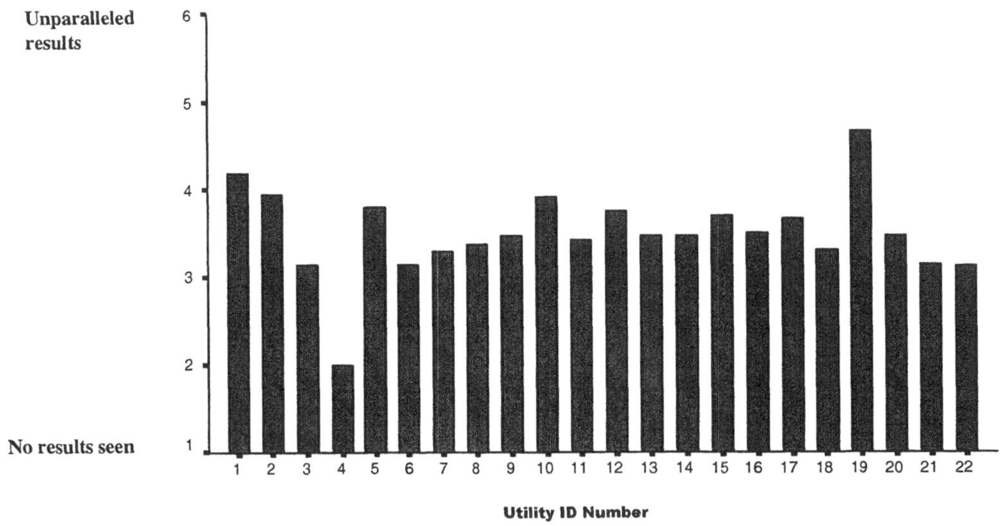

Overall Statistics			
Mean	3.5	Minimum	1
Median	4.0	Maximum	6
Std. Dev.	1.1	Population	22

Category: Participation/Style

Q54 - This utility uses team problem solving routinely across all utility functions to achieve its performance improvement goals

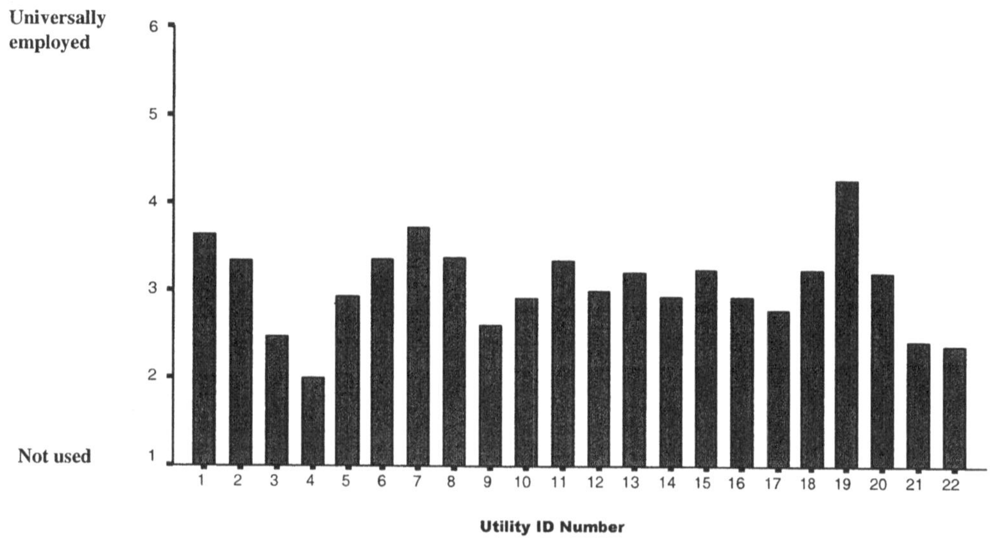

Overall Statistics			
Mean	3.1	Minimum	1
Median	3.0	Maximum	6
Std. Dev.	1.3	Population	22

Category: Participation/Style

Q55 - This utility uses a team-based approach in business processes and work practices

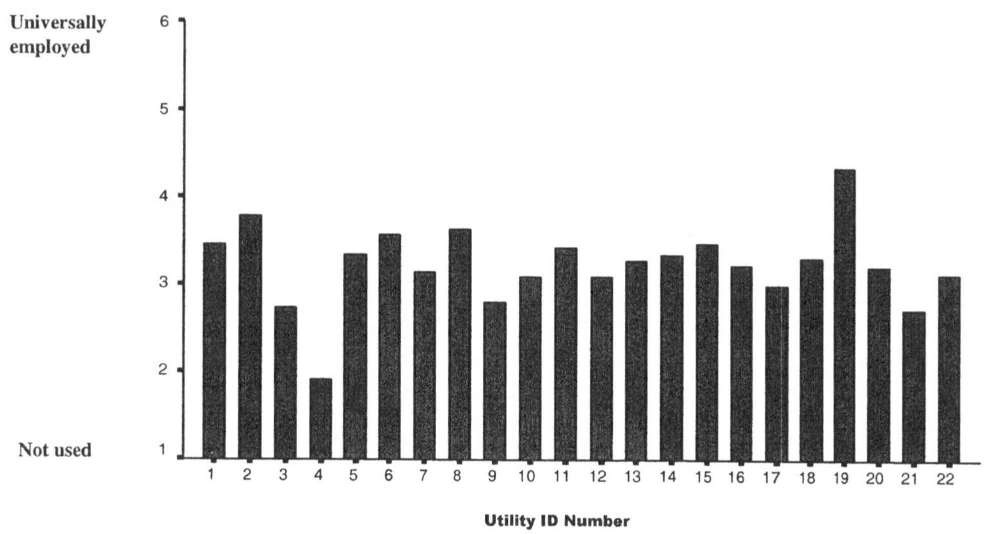

Overall Statistics			
Mean	3.2	Minimum	1
Median	3.0	Maximum	6
Std. Dev.	1.2	Population	22

Category: Participation/Style
Q61 - Labor and management at this utility meet frequently regarding planned initiatives for change in addition to meetings dealing with grievances or contract negotiations

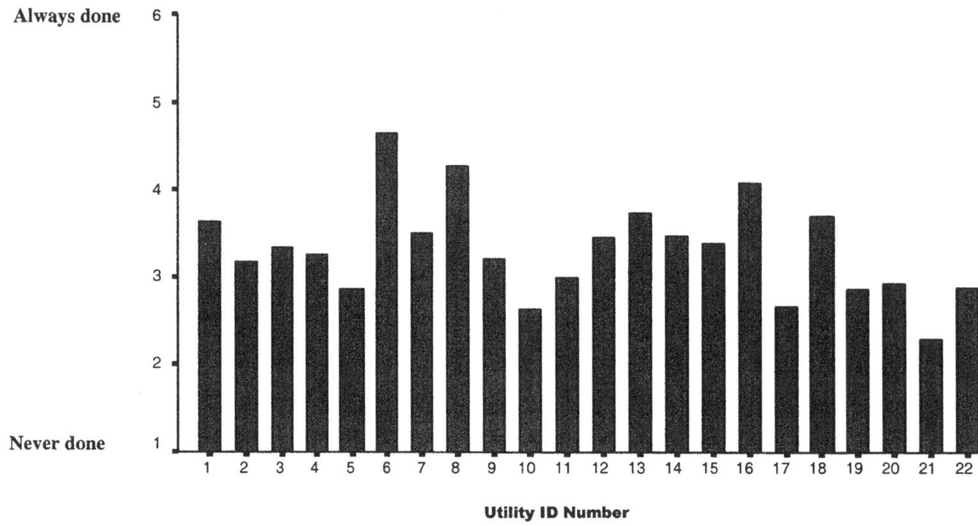

Overall Statistics			
Mean	3.3	Minimum	1
Median	3.0	Maximum	6
Std. Dev.	1.3	Population	22

Category: Participation/Style
Q73 - Work practice redesign processes at this utility use a team-based approach

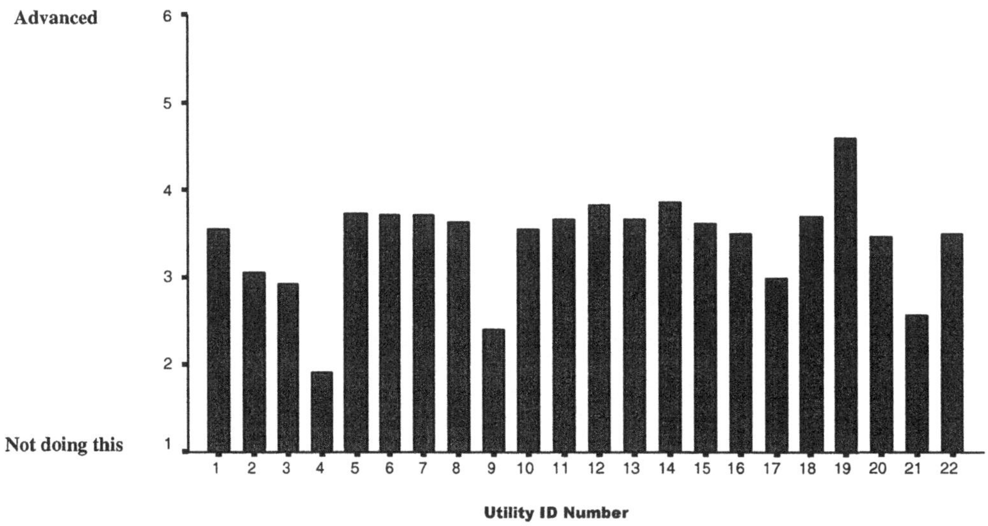

Overall Statistics			
Mean	3.4	Minimum	1
Median	3.8	Maximum	6
Std. Dev.	1.3	Population	22

Category: Communications
Q18 - This utility effectively communicates about change

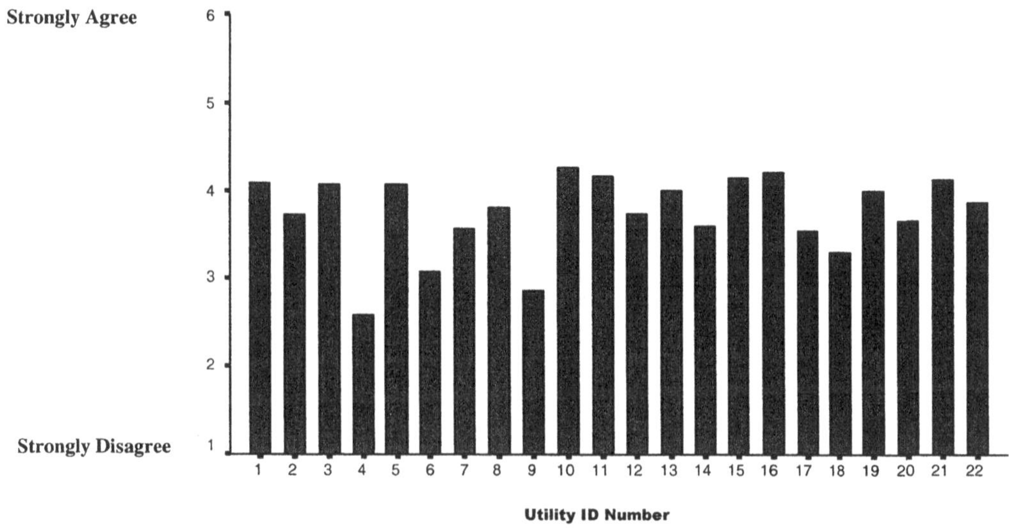

Overall Statistics			
Mean	3.8	Minimum	1
Median	4.0	Maximum	6
Std. Dev.	1.1	Population	22

Category: Communications
Q19 - There is an effective process for employees to offer input and receive feedback on issues that concern them

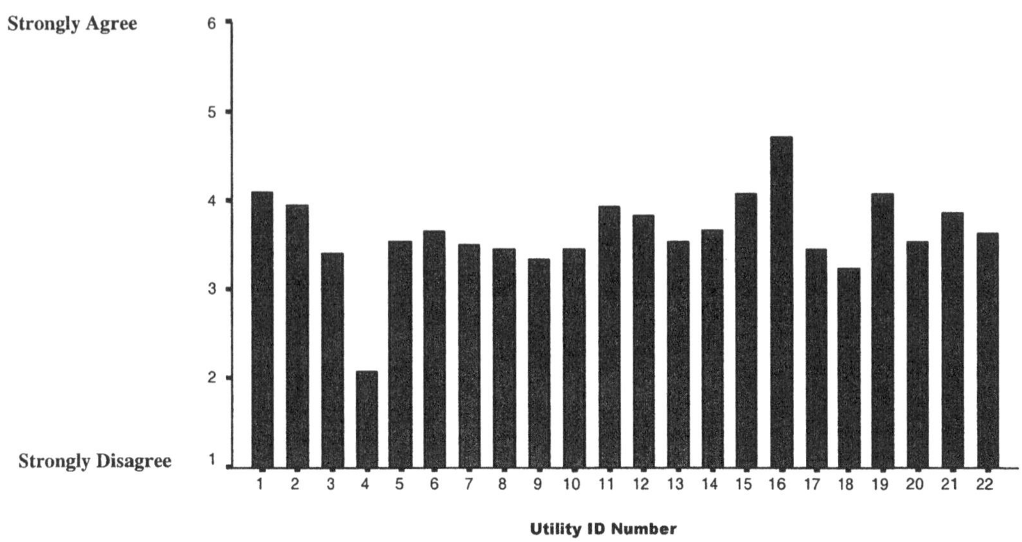

Overall Statistics			
Mean	3.6	Minimum	1
Median	4.0	Maximum	6
Std. Dev.	1.2	Population	22

Category: Communications
Q20 - This utility has effective processes to respond to the voice of customers and other stakeholders

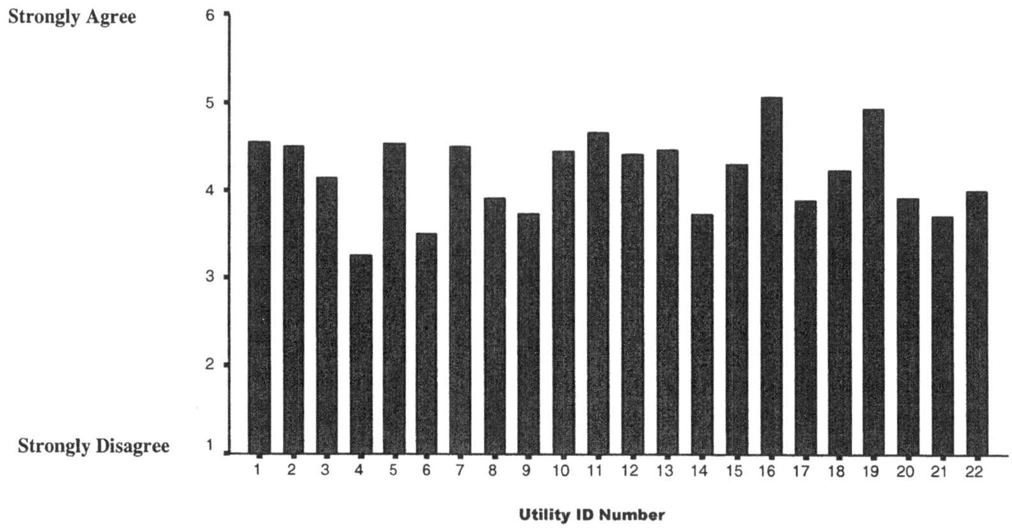

Overall Statistics			
Mean	4.2	Minimum	1
Median	4.0	Maximum	6
Std. Dev.	1.0	Population	22

Category: Communications
Q21 - Managers and supervisors at this utility consistently provide feedback to individual employees about suggestions or issues that concern them

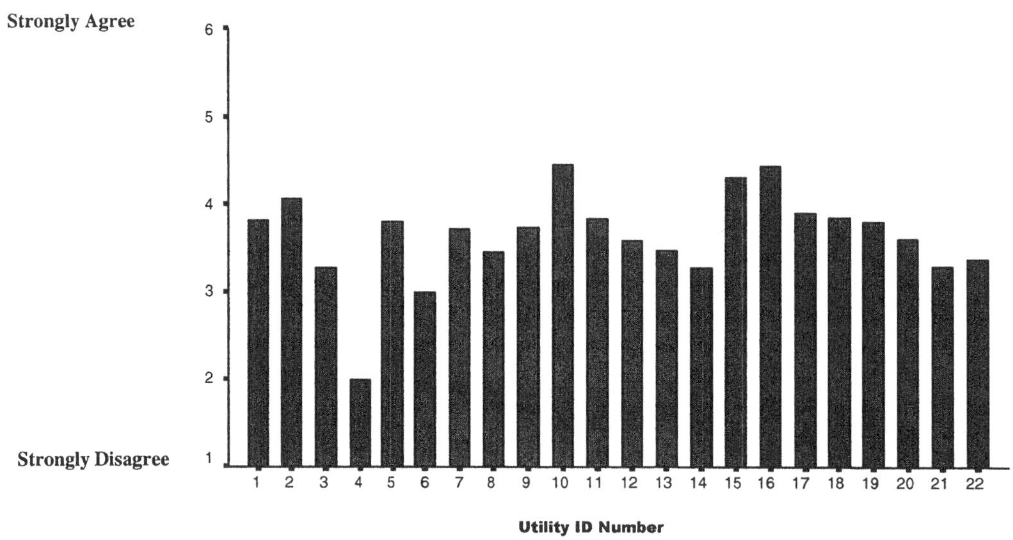

Overall Statistics			
Mean	3.6	Minimum	1
Median	4.0	Maximum	6
Std. Dev.	1.2	Population	22

Category: Communications

Q49 - Communications regarding long-term goals within this utility have been effective in yielding desired results

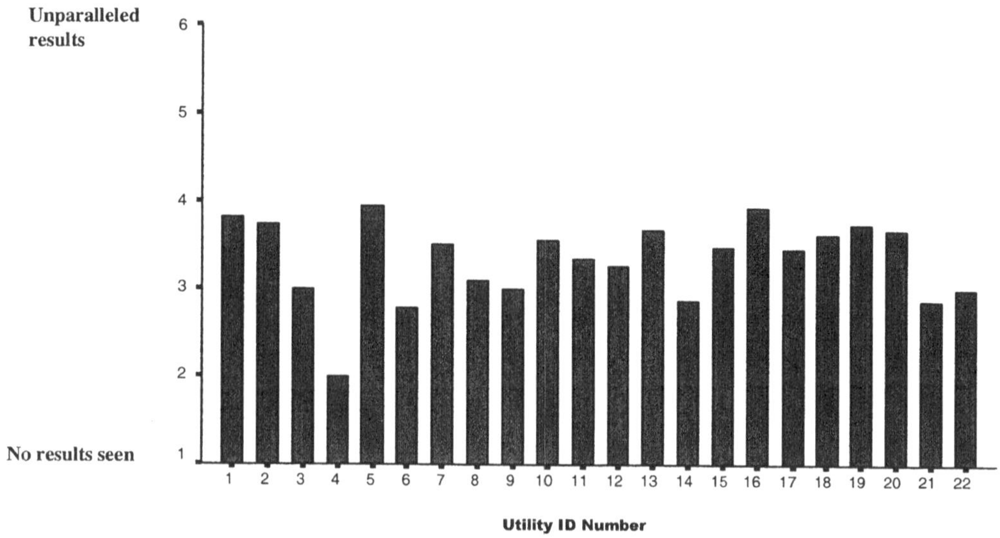

Overall Statistics			
Mean	3.3	Minimum	1
Median	3.1	Maximum	6
Std. Dev.	1.0	Population	22

Category: Communications

Q69 - This utility has a formal, published communications plan/system

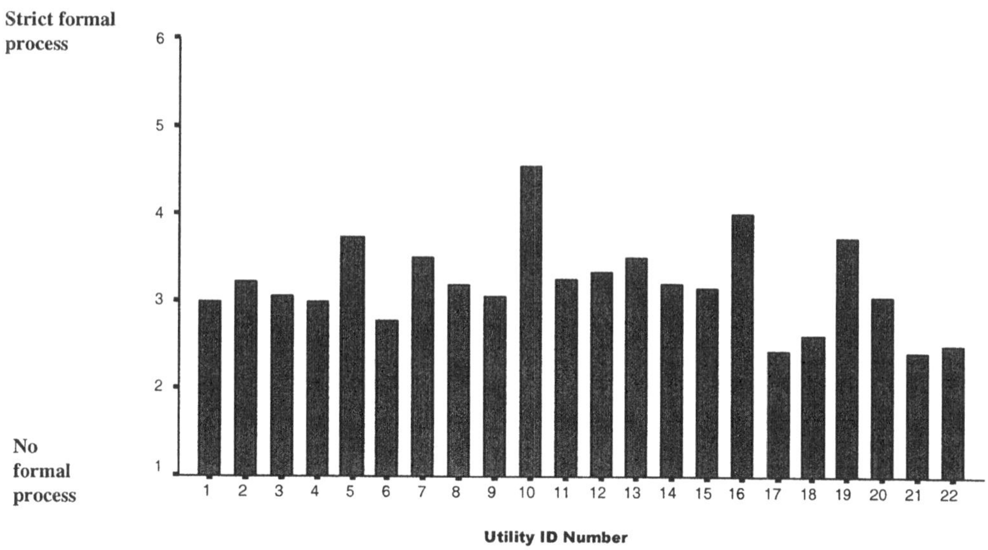

Overall Statistics			
Mean	3.2	Minimum	1
Median	3.0	Maximum	6
Std. Dev.	1.3	Population	22

Category: Communications
Q83 - Using formal communications well is important to this utility

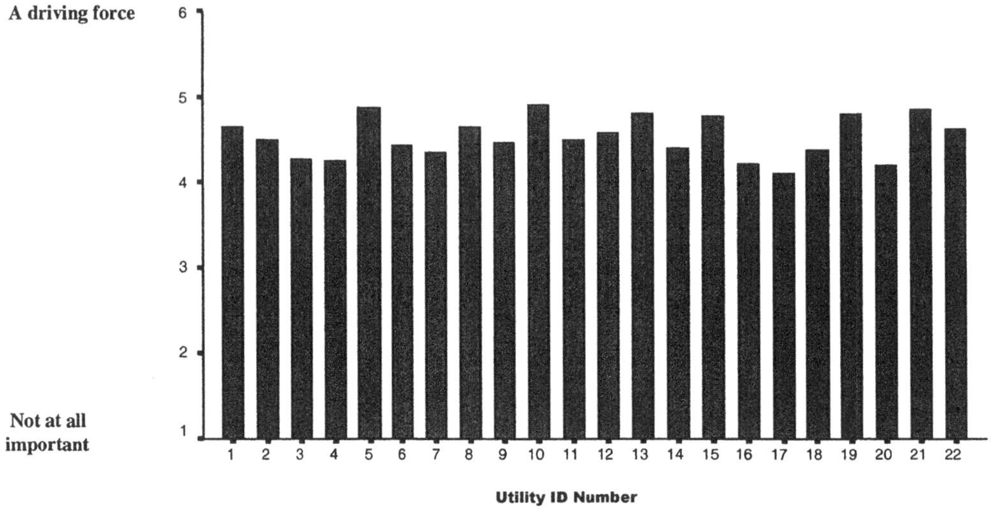

Overall Statistics			
Mean	4.5	Minimum	1
Median	5.0	Maximum	6
Std. Dev.	1.0	Population	22

Category: Communications
Q84 - It is important at this utility to capture and share lessons learned in the change process

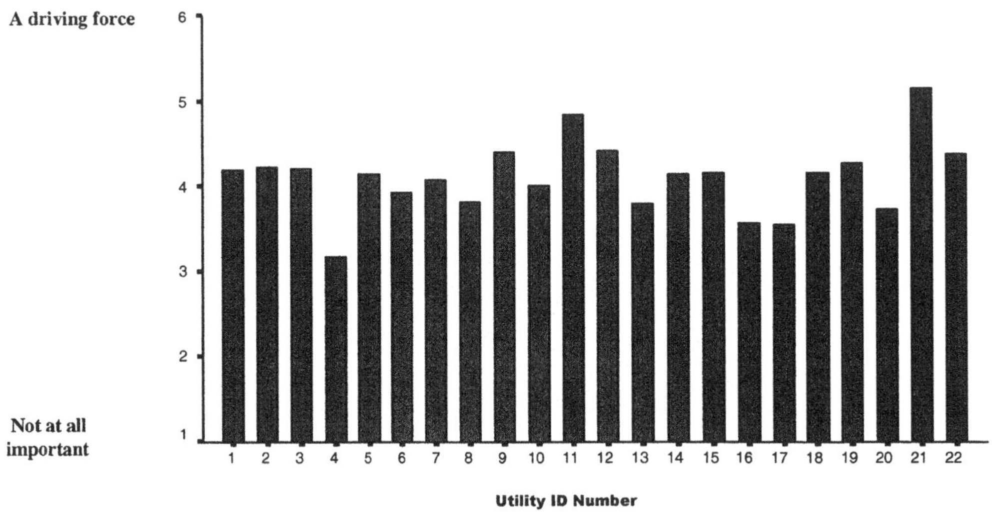

Overall Statistics			
Mean	4.1	Minimum	1
Median	4.0	Maximum	6
Std. Dev.	1.3	Population	22

Category: Measures

Q22 - I believe that there is a need for measuring business process or work practice as a basis for utility performance improvement

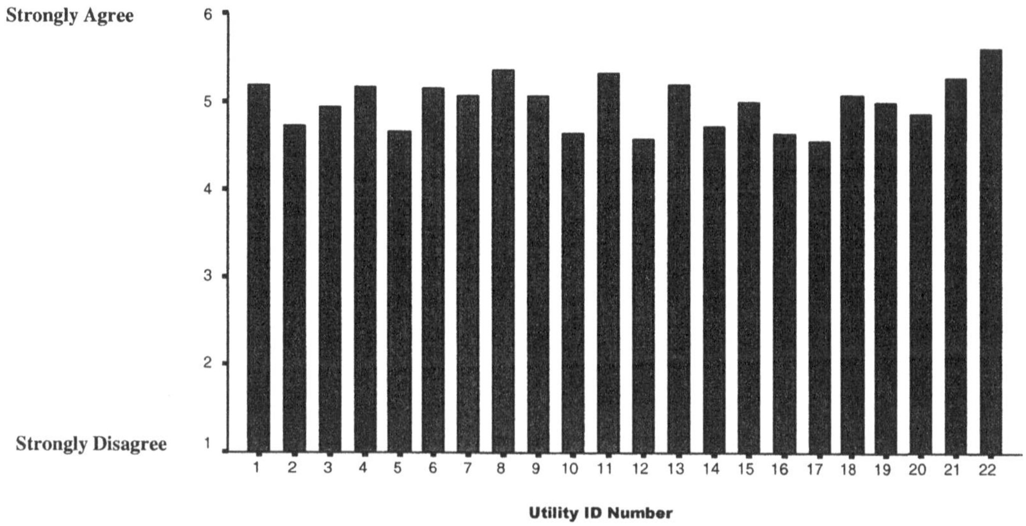

Overall Statistics			
Mean	5.0	Minimum	1
Median	5.0	Maximum	6
Std. Dev.	.9	Population	22

Category: Measures

Q23 - I believe that there is a need for measuring individual performance as a basis for improvement

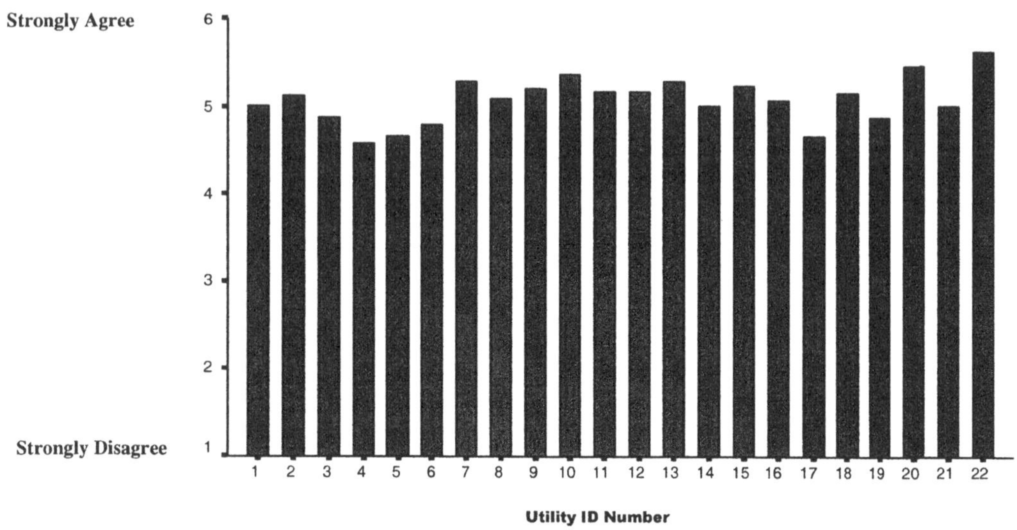

Overall Statistics			
Mean	5.1	Minimum	2
Median	5.0	Maximum	6
Std. Dev.	.8	Population	22

Category: Measures
Q24 - Measurement of business processes and work practices at this utility has contributed to improved business performance and increased customer satisfaction

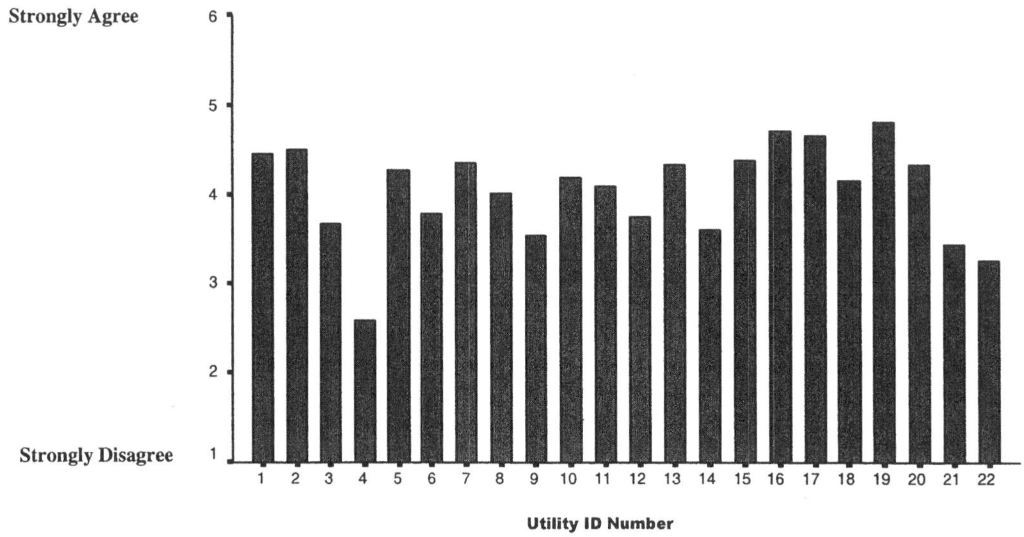

Overall Statistics			
Mean	4.0	Minimum	1
Median	4.0	Maximum	6
Std. Dev.	1.2	Population	22

Category: Measures
Q25 - Individual performance measures in this utility are effective in balancing efficiency, quality, effectiveness, and customer service

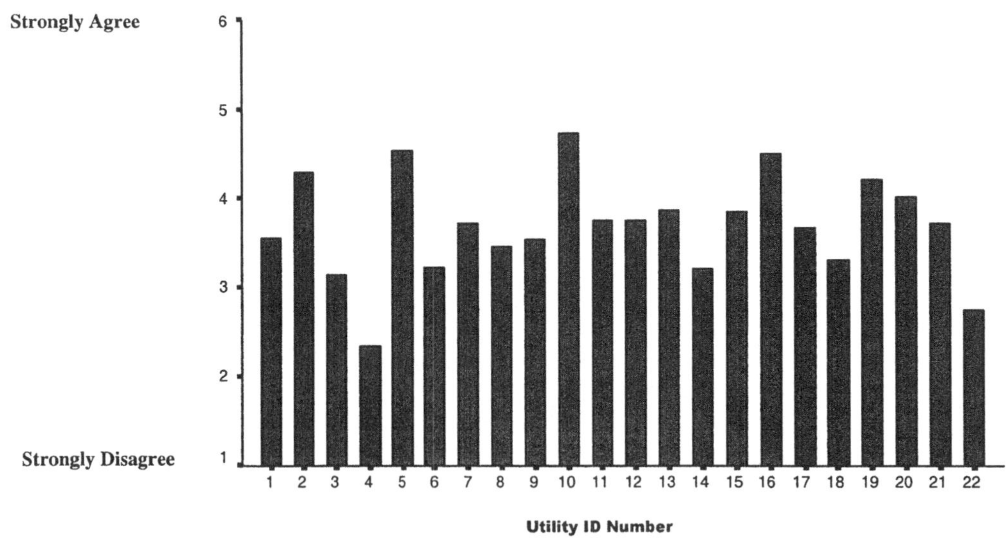

Overall Statistics			
Mean	3.7	Minimum	1
Median	4.0	Maximum	6
Std. Dev.	1.2	Population	22

Category: Measures
Q26 - Within the past 12 months, I have been significantly impacted by major change efforts in this utility

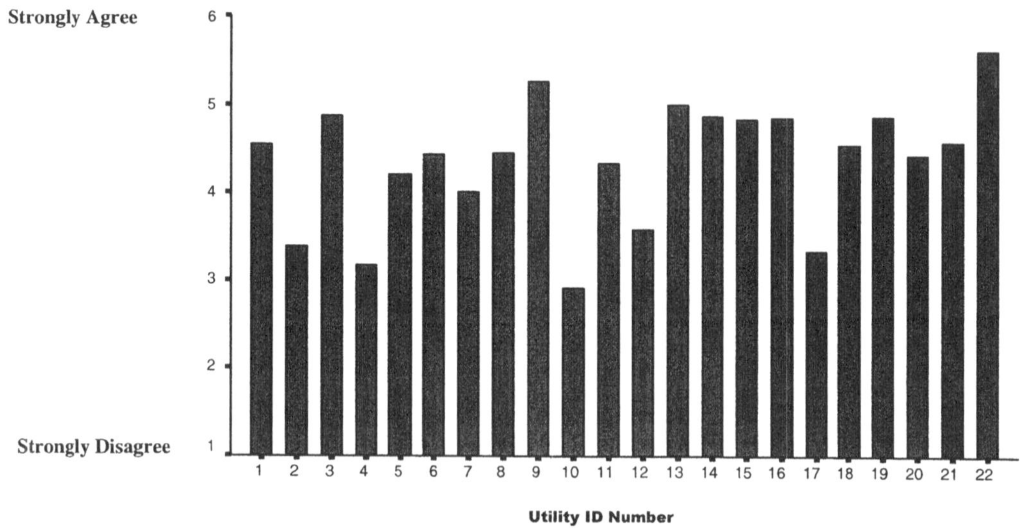

Overall Statistics			
Mean	4.4	Minimum	1
Median	5.0	Maximum	6
Std. Dev.	1.5	Population	22

Category: Measures
Q50 - I see evidence that this utility is becoming more efficient and competitive

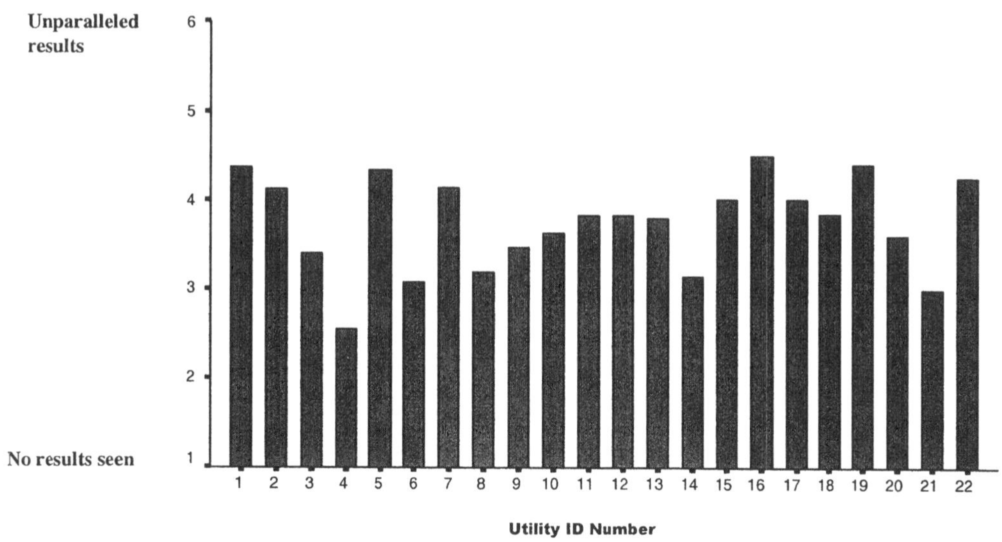

Overall Statistics			
Mean	3.8	Minimum	1
Median	4.0	Maximum	6
Std. Dev.	1.1	Population	22

Category: Measures
Q53 - This utility uses specific performance measures to improve individual performance

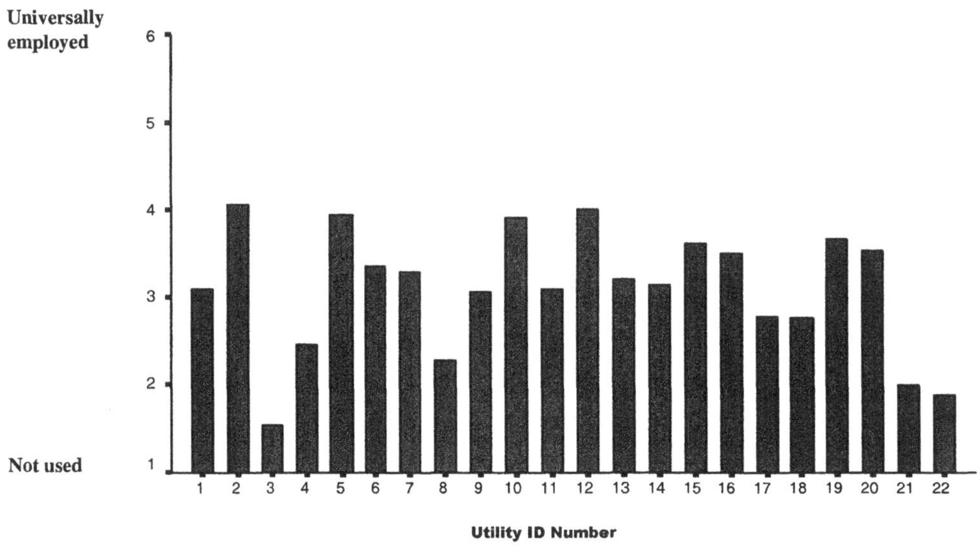

Overall Statistics			
Mean	3.1	Minimum	1
Median	3.0	Maximum	6
Std. Dev.	1.4	Population	22

Category: Measures
Q56 - This utility includes individual performance measures of customer service

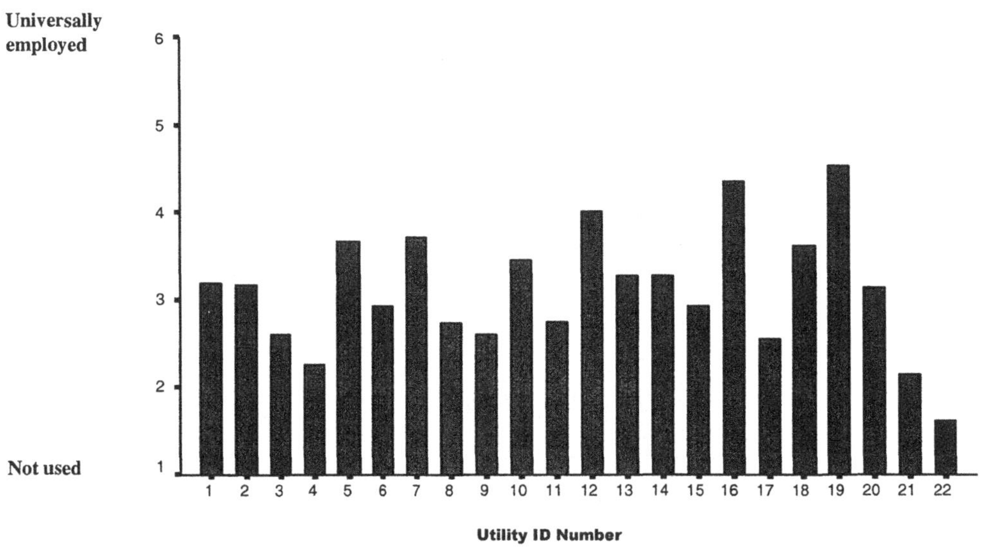

Overall Statistics			
Mean	3.1	Minimum	1
Median	3.0	Maximum	6
Std. Dev.	1.4	Population	22

Category: Measures

Q57 - This utility uses specific performance measures to improve business processes and work practices

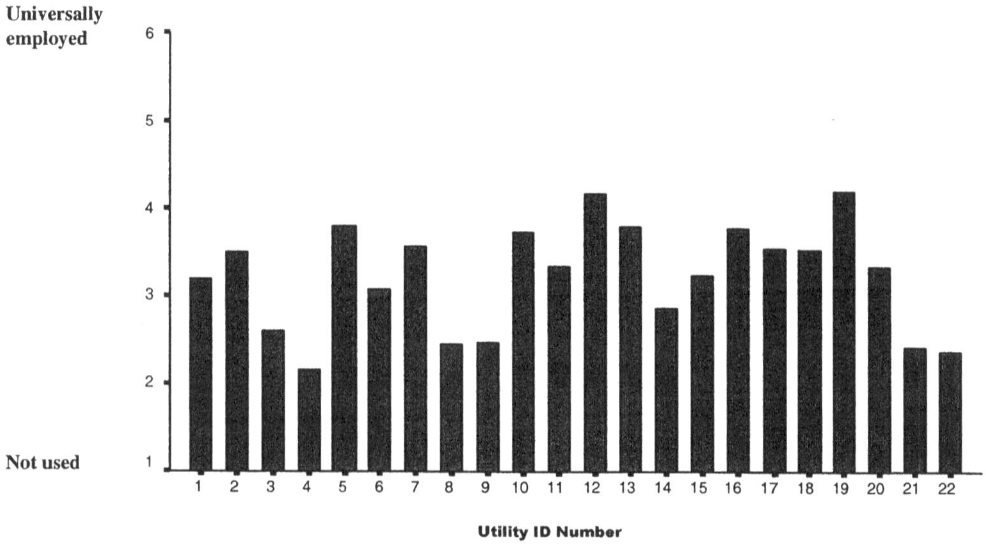

Overall Statistics			
Mean	3.2	Minimum	1
Median	3.0	Maximum	6
Std. Dev.	1.3	Population	22

Category: Measures
Q85 - It is important for this utility to compare its performance with other utilities

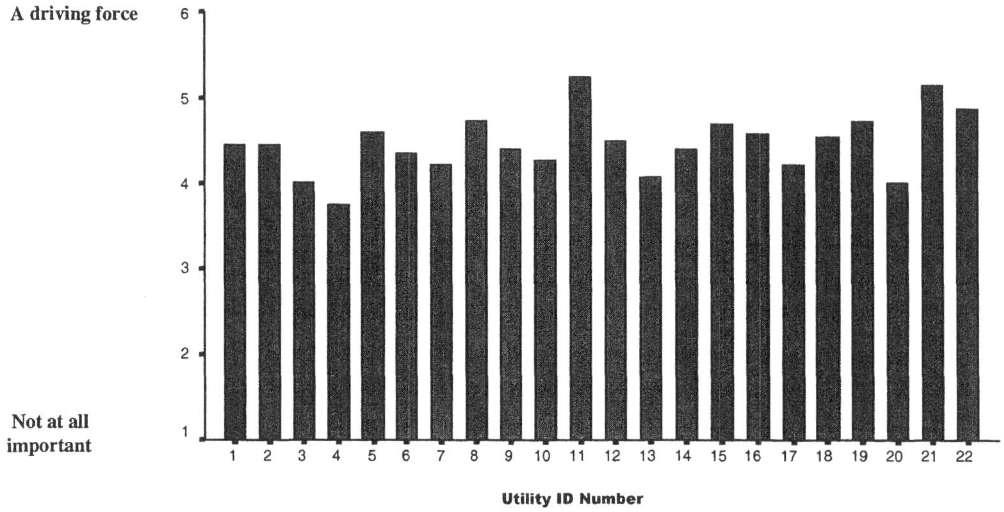

Overall Statistics			
Mean	4.5	Minimum	1
Median	5.0	Maximum	6
Std. Dev.	1.2	Population	22

Category: Workforce Flexibility
Q27 - Cross-training of employees in this utility has been effective in achieving a multi-skilled workforce

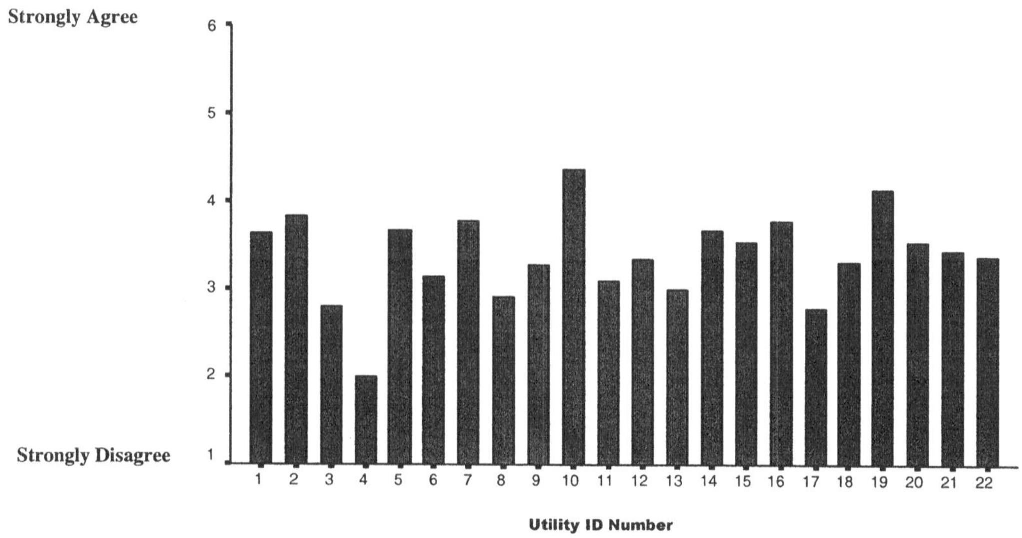

Overall Statistics			
Mean	3.4	Minimum	1
Median	3.5	Maximum	6
Std. Dev.	1.3	Population	22

Category: Workforce Flexibility
Q28 - The easiest way to get promoted in this utility is to become more specialized in a particular skill area

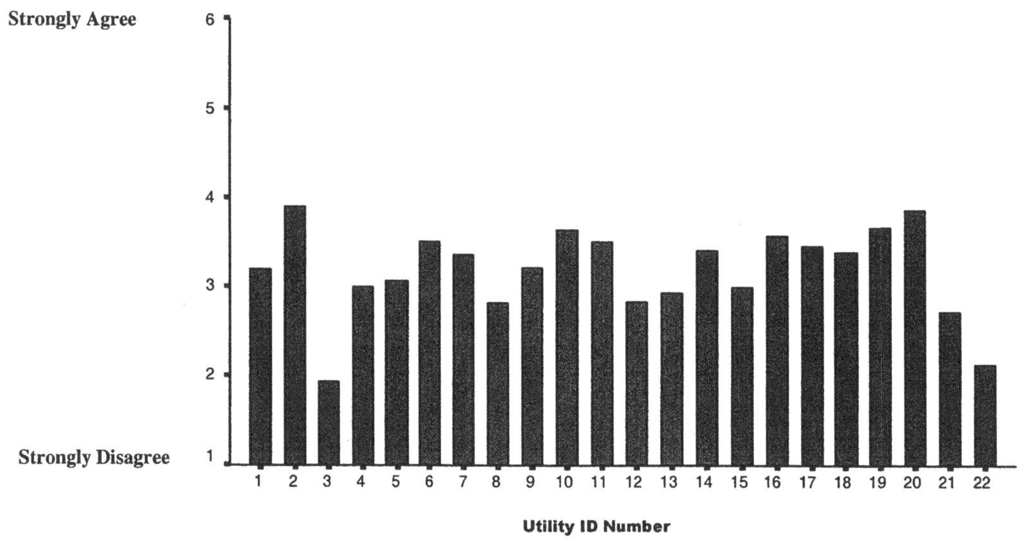

Overall Statistics			
Mean	3.2	Minimum	1
Median	3.0	Maximum	6
Std. Dev.	1.3	Population	22

Category: Workforce Flexibility

Q29 - I am confident that this utility will invest in appropriate tools and training as they are needed to make major changes to improve business performance and customer satisfaction

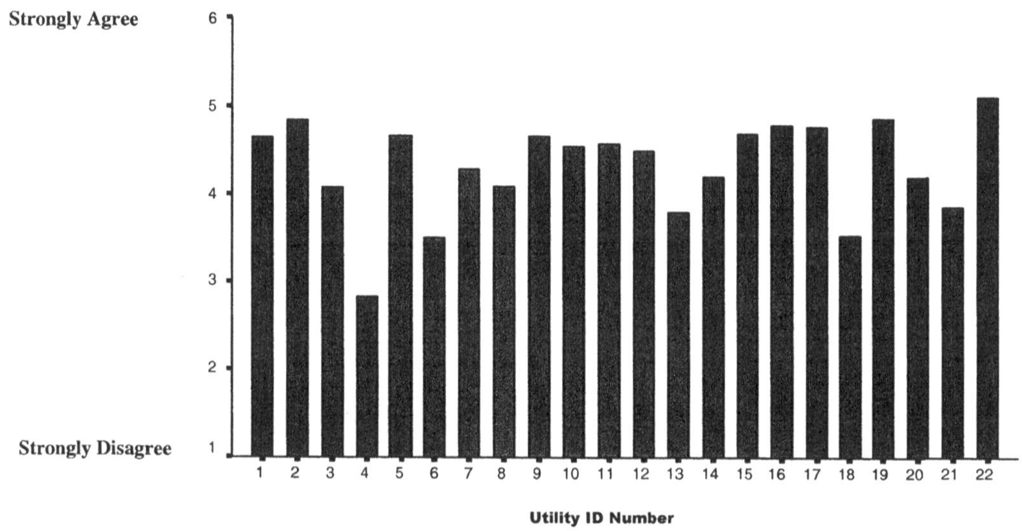

Overall Statistics			
Mean	4.3	Minimum	1
Median	5.0	Maximum	6
Std. Dev.	1.2	Population	22

Category: Workforce Flexibility

Q30 - This utility has achieved greater flexibility in work performance through the use of a multi-skilled workforce

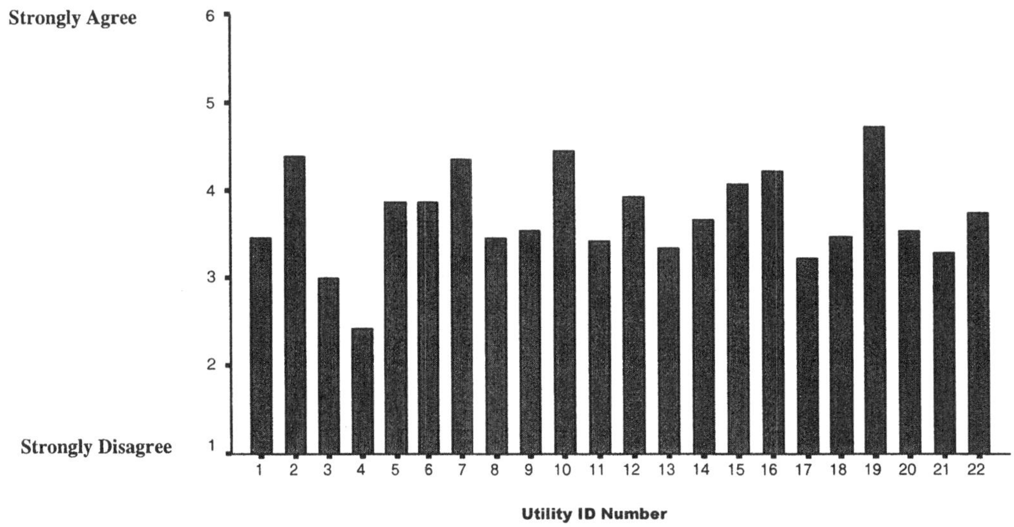

Overall Statistics			
Mean	3.7	Minimum	1
Median	4.0	Maximum	6
Std. Dev.	1.2	Population	22

Category: Workforce Flexibility
Q58 - Diversity goals are considered in this utility's initiatives for cross-training

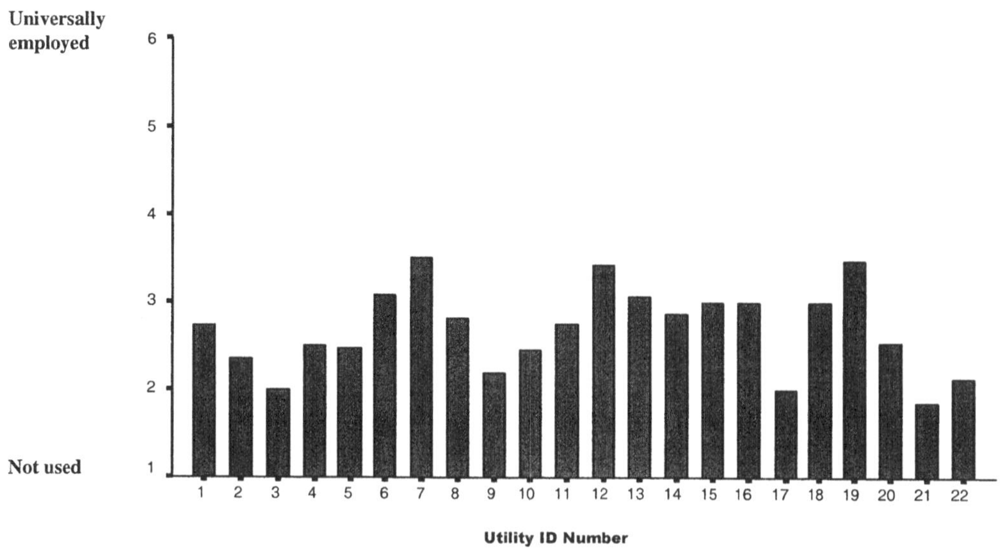

Overall Statistics			
Mean	2.7	Minimum	1
Median	3.0	Maximum	6
Std. Dev.	1.4	Population	22

Category: Workforce Flexibility
Q59 - This utility uses a multi-skilled workforce

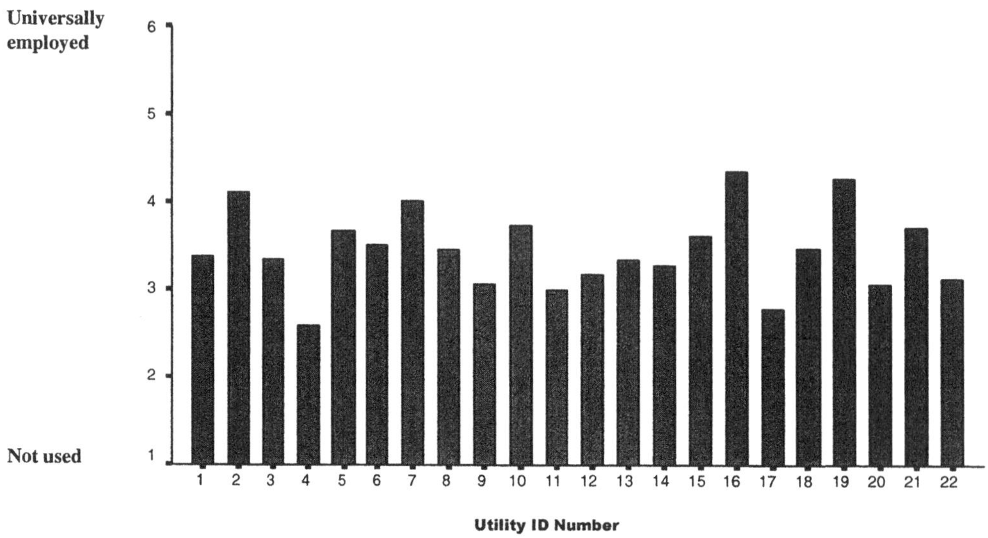

Overall Statistics			
Mean	3.5	Minimum	1
Median	3.0	Maximum	6
Std. Dev.	1.3	Population	22

Category: Workforce Flexibility
Q63 - This utility has formal succession planning for all key management positions

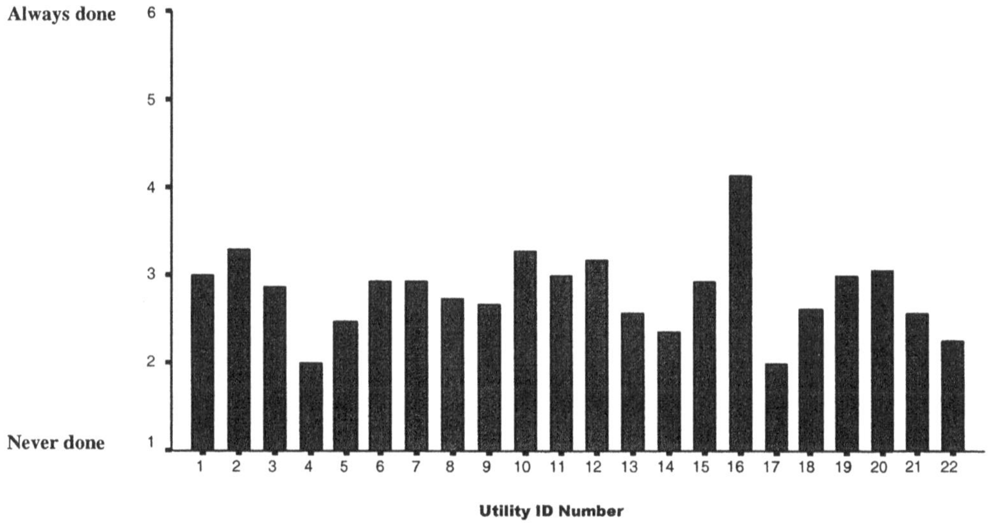

Overall Statistics			
Mean	2.8	Minimum	1
Median	3.0	Maximum	6
Std. Dev.	1.3	Population	22

Category: Workforce Flexibility
Q64 - This utility has complete, up-to-date profiles for all employees

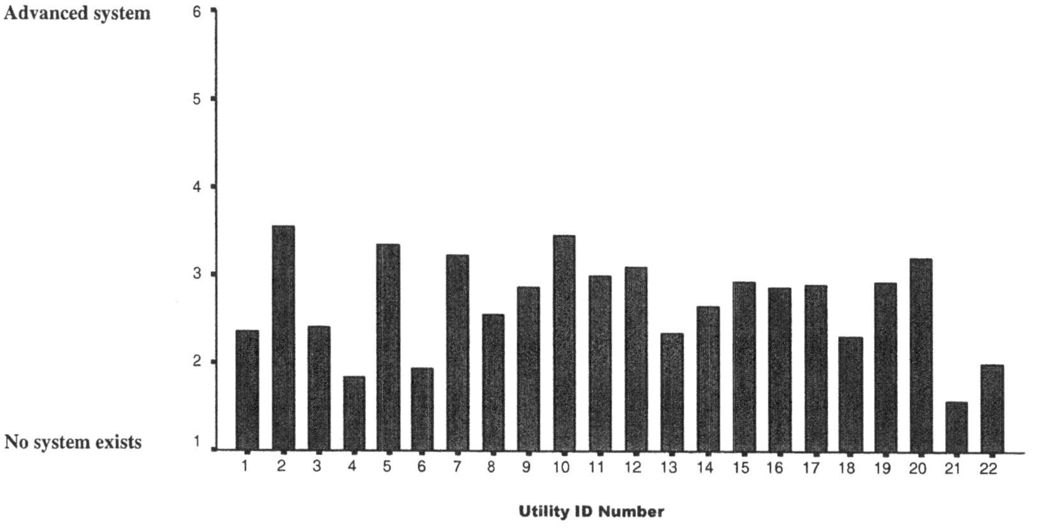

Overall Statistics			
Mean	2.7	Minimum	1
Median	2.8	Maximum	6
Std. Dev.	1.4	Population	22

Category: Workforce Flexibility

Q66 - This utility practices employee cross-training, certification and new skills development through a regular budgeted investment process

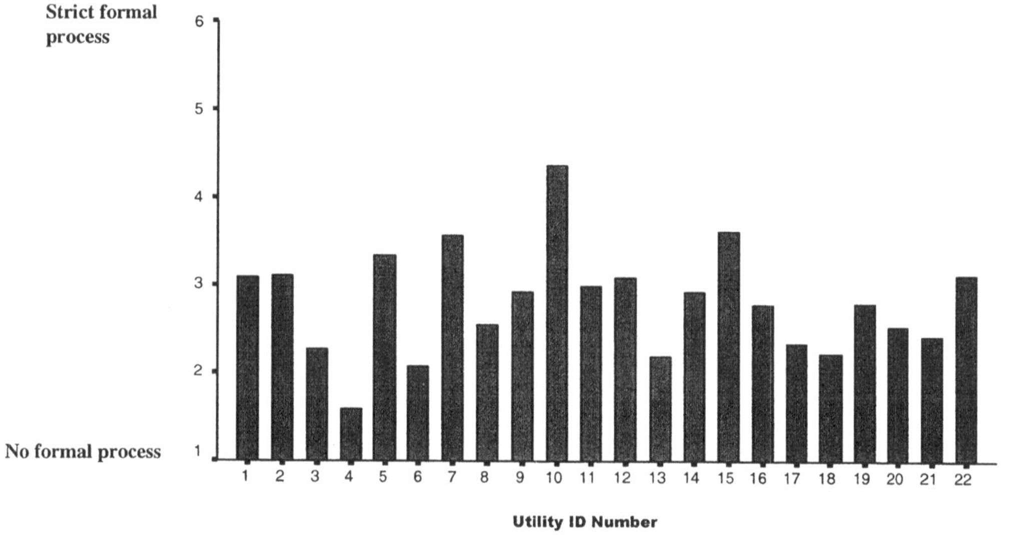

Overall Statistics			
Mean	2.8	Minimum	1
Median	3.0	Maximum	6
Std. Dev.	1.3	Population	22

Category: Measures
Q77 - Management personnel at this utility receive _____amount of formal training annually

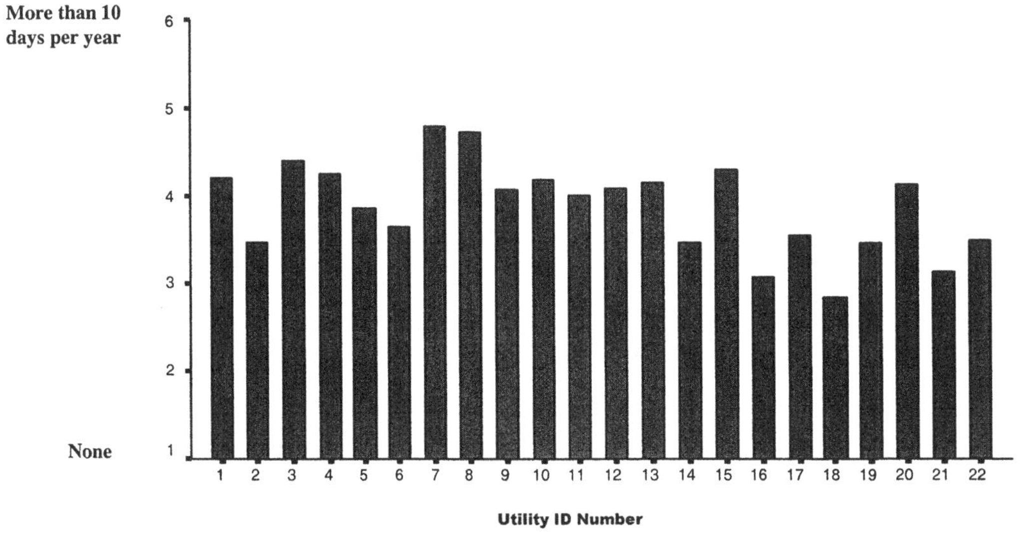

Overall Statistics			
Mean	3.9	Minimum	1
Median	4.0	Maximum	6
Std. Dev.	1.3	Population	22

Category: Measures

Q78 - Employees at this utility receive _____ amount of formal training annually

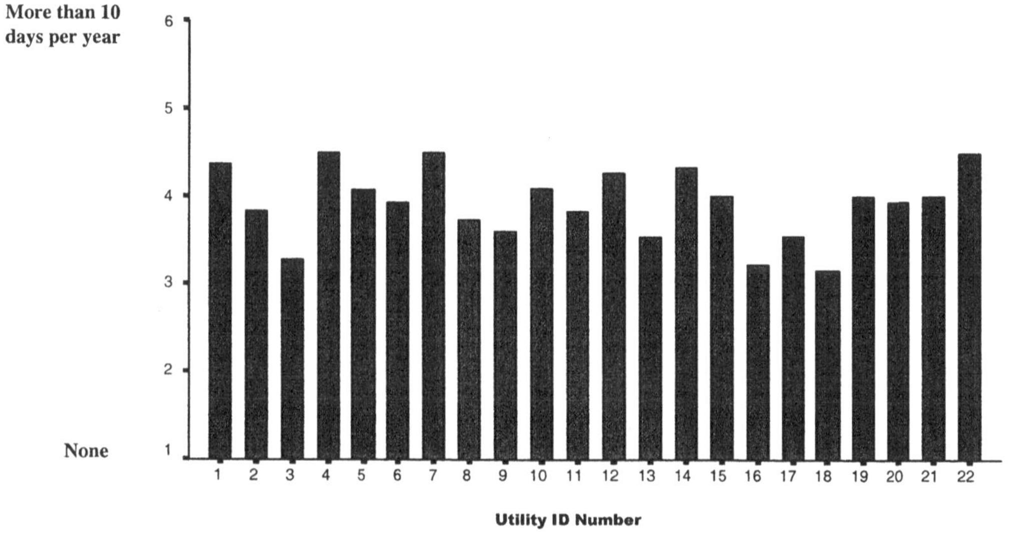

Overall Statistics			
Mean	3.9	Minimum	1
Median	4.0	Maximum	6
Std. Dev.	1.1	Population	22

Category: Compensation/Rewards

Q31 - This utility has effective individual performance appraisal and/or review process

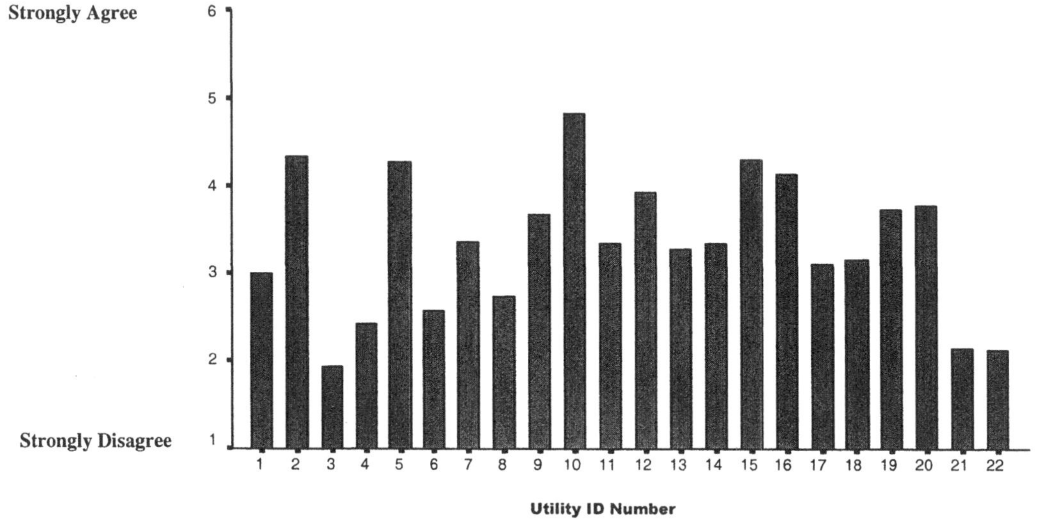

Overall Statistics			
Mean	3.3	Minimum	1
Median	4.0	Maximum	6
Std. Dev.	1.5	Population	22

Category: Compensation/Rewards

Q32 - I believe that employees who stay with this utility should be rewarded with cost of living based pay increases and job security regardless of market-based forces or this utility's performance

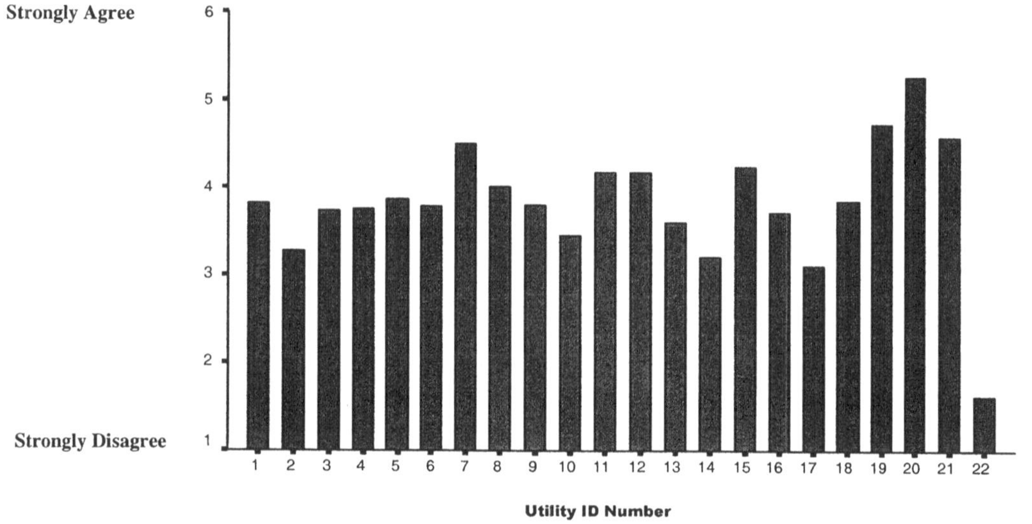

Overall Statistics			
Mean	3.8	Minimum	1
Median	4.0	Maximum	6
Std. Dev.	1.6	Population	22

Category: Compensation/Rewards
Q33 - Individual annual compensation should be impacted by this utility's business performance

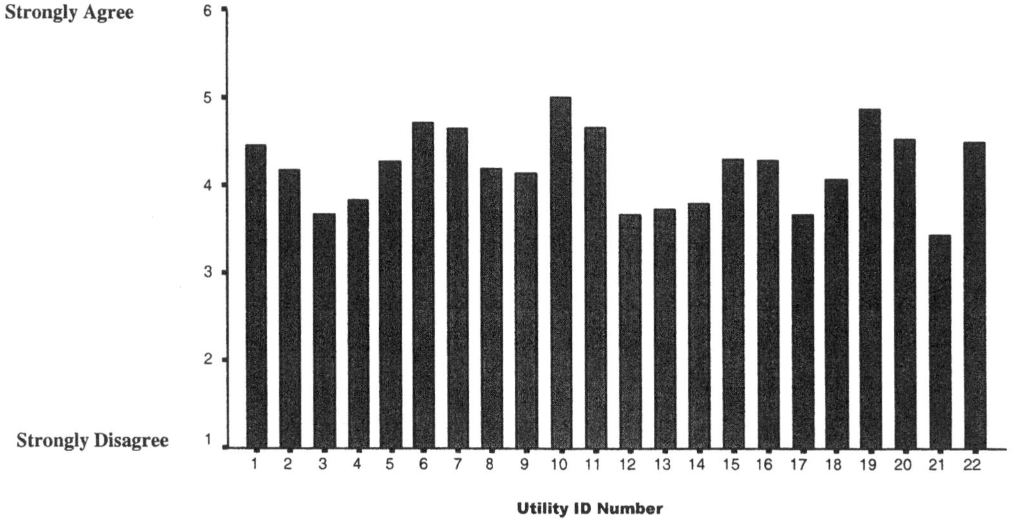

Overall Statistics			
Mean	4.2	Minimum	1
Median	4.0	Maximum	6
Std. Dev.	1.2	Population	22

Category: Compensation/Rewards
Q34 - Individual annual compensation should be impacted by achievement of team performance goals and objectives

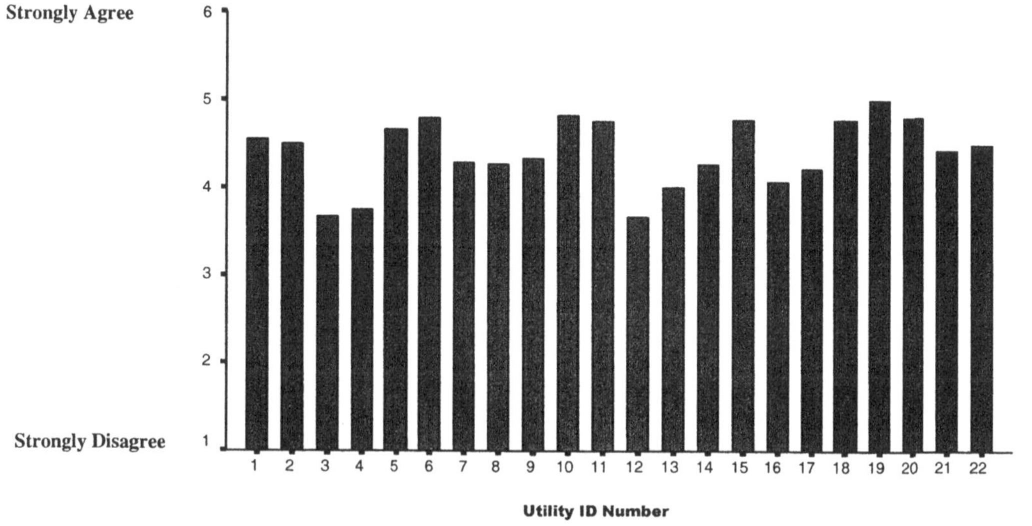

Overall Statistics			
Mean	4.4	Minimum	1
Median	4.8	Maximum	6
Std. Dev.	1.1	Population	22

Category: Compensation/Rewards

Q60 - Employees at this utility are rewarded for new ideas that improve business processes and work practices

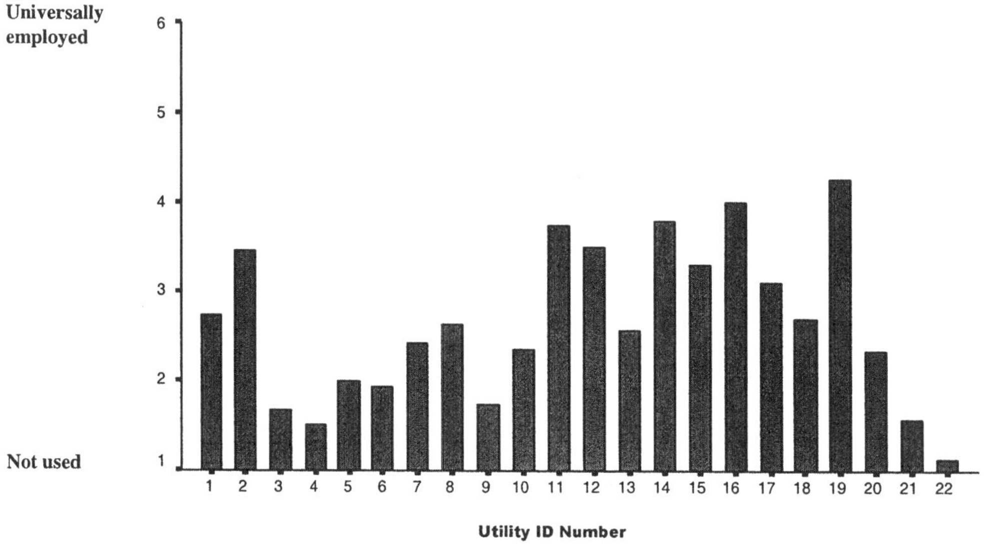

Overall Statistics

Mean	2.7	Minimum	1
Median	3.0	Maximum	6
Std. Dev.	1.5	Population	22

Category: Compensation/Rewards
Q67 - Individual performance evaluation processes include specific quantitative and qualitative criteria

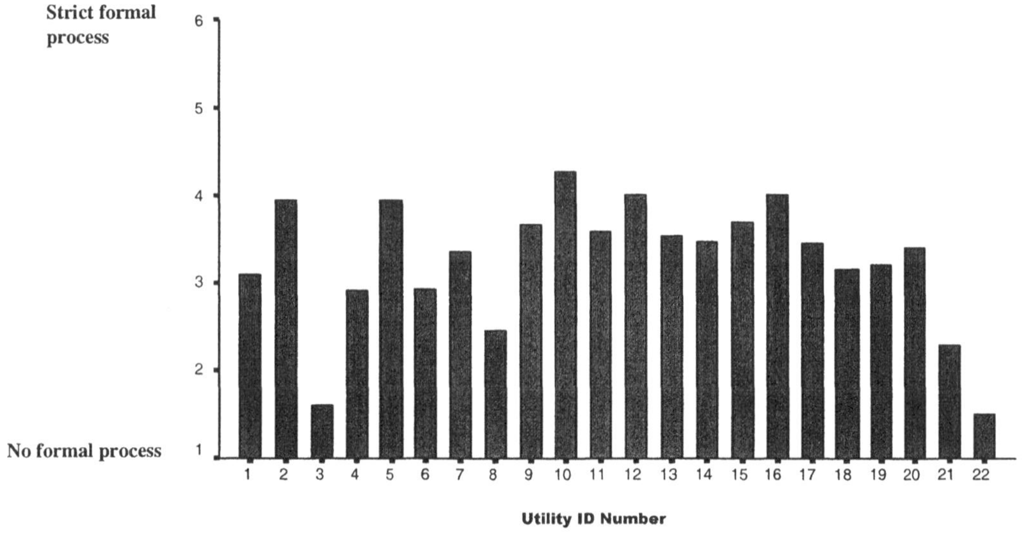

Overall Statistics			
Mean	3.2	Minimum	1
Median	3.3	Maximum	6
Std. Dev.	1.3	Population	22

Category: Compensation/Rewards

Q68 - A formal system of rewards exists at this utility for thinking creatively to improve business processes and work practices

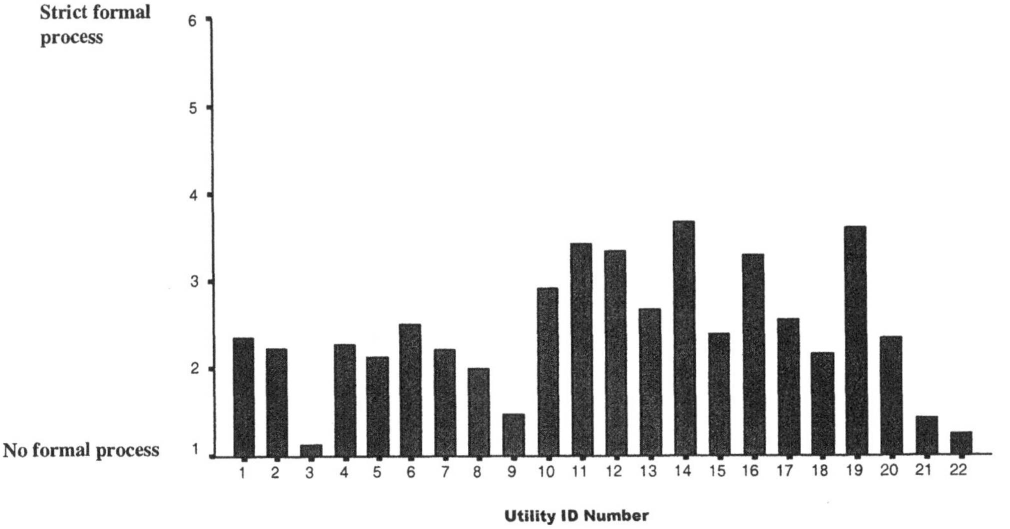

Overall Statistics			
Mean	2.4	Minimum	1
Median	2.0	Maximum	6
Std. Dev.	1.4	Population	22

Category: Compensation/Rewards
Q72 - This utility has an effective employee non-monetary recognition/awards program

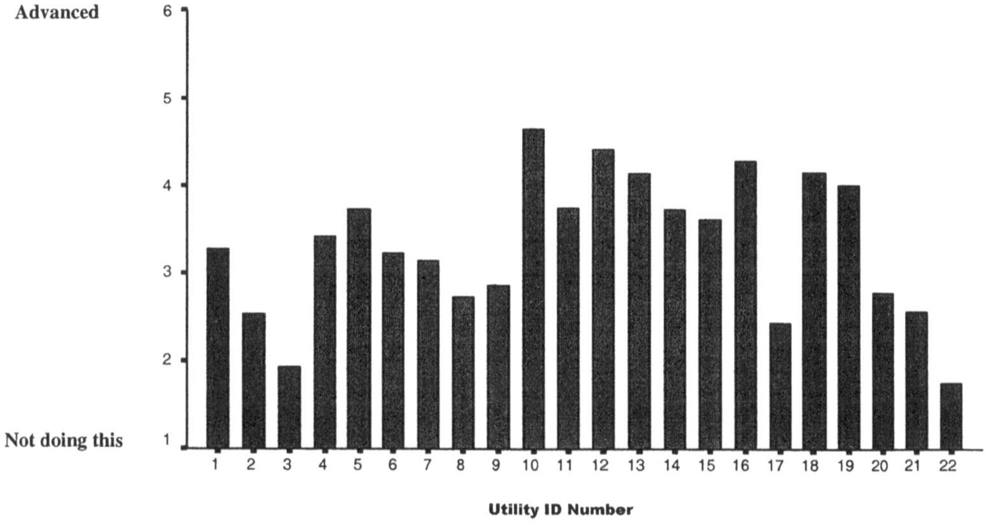

Overall Statistics			
Mean	3.3	Minimum	1
Median	4.0	Maximum	6
Std. Dev.	1.4	Population	22

Category: Policies/Procedures/Tools
Q35 - This utility has effective processes in place for identifying critical hiring needs

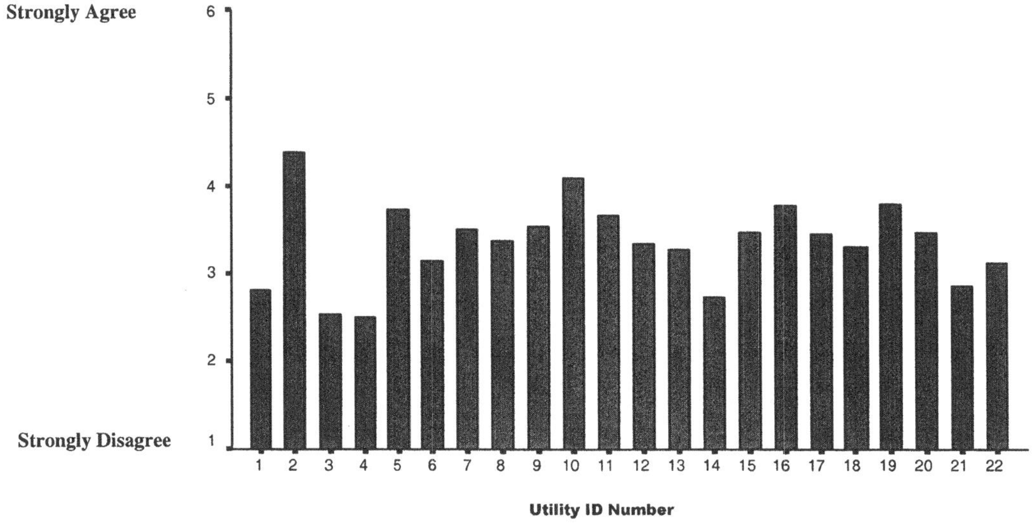

Overall Statistics			
Mean	3.4	Minimum	1
Median	3.0	Maximum	6
Std. Dev.	1.3	Population	22

Category: Policies/Procedures/Tools
Q36 - This utility has effective processes in place for hiring competent new staff

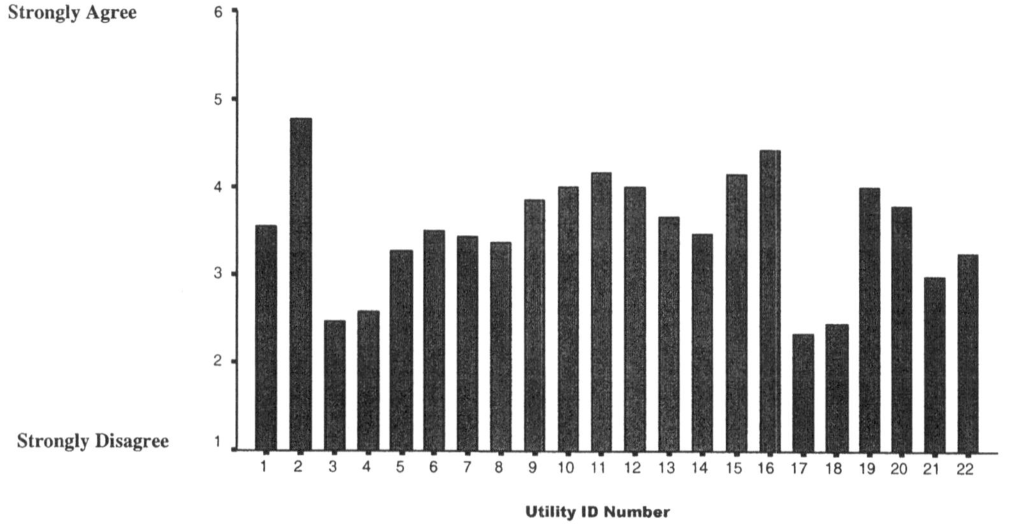

Overall Statistics			
Mean	3.5	Minimum	1
Median	4.0	Maximum	6
Std. Dev.	1.4	Population	22

Category: Policies/Procedures/Tools
Q37 - This utility effectively deals with personnel issues

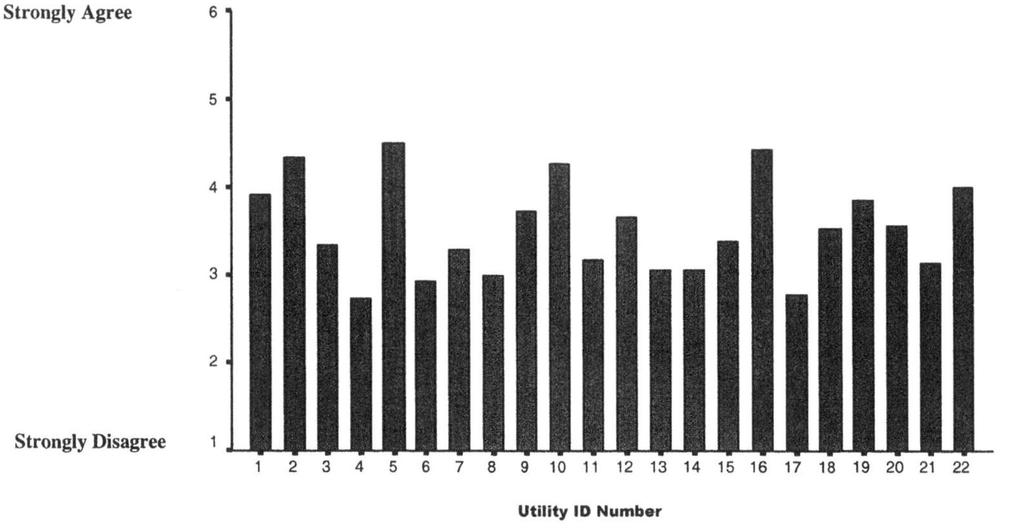

Overall Statistics			
Mean	3.5	Minimum	1
Median	4.0	Maximum	6
Std. Dev.	1.3	Population	22

Category: Policies/Procedures/Tools
Q52 - This utility emphasizes employee wellness by providing assistance

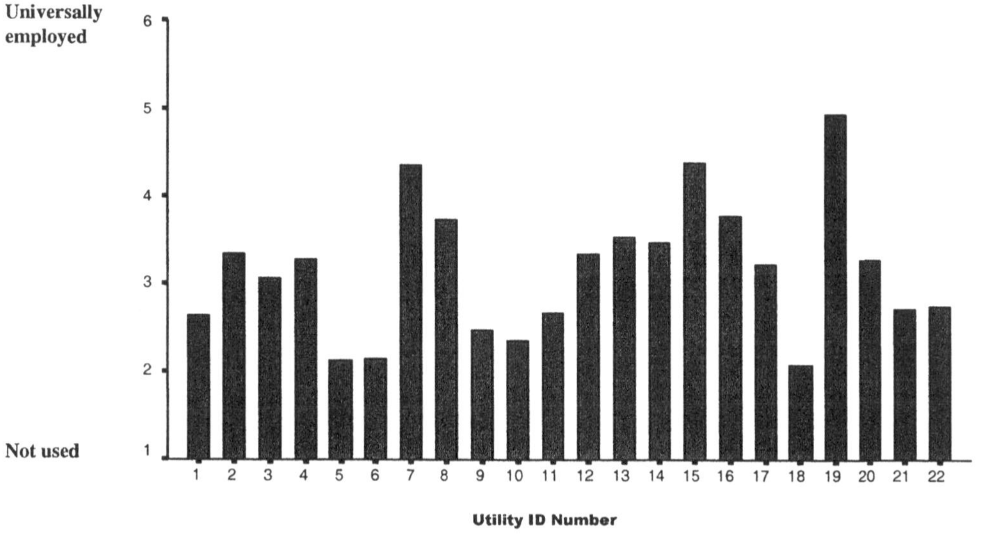

Overall Statistics			
Mean	3.2	Minimum	1
Median	3.0	Maximum	6
Std. Dev.	1.5	Population	22

Category: Policies/Procedures/Tools
Q65 - This utility uses information technology for human resource management

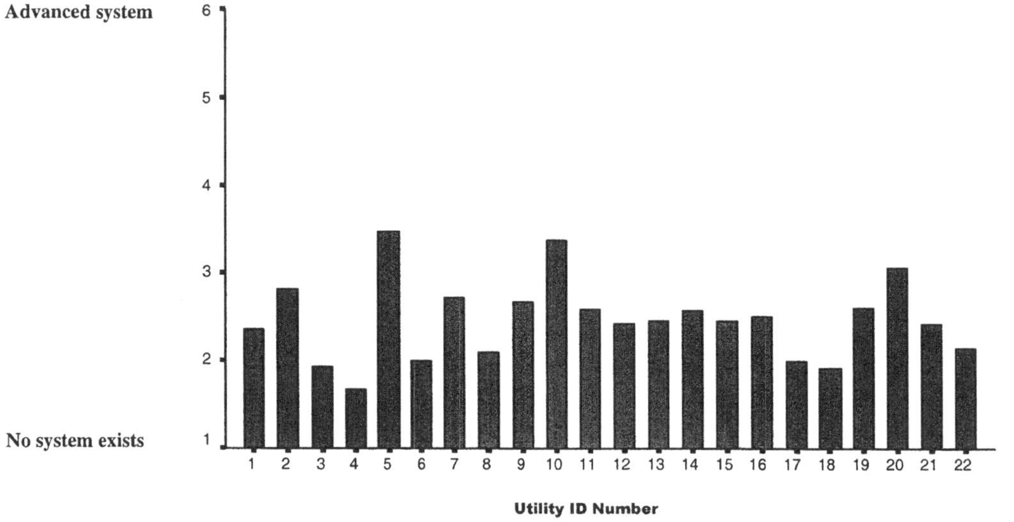

Overall Statistics			
Mean	2.5	Minimum	1
Median	2.0	Maximum	6
Std. Dev.	1.3	Population	22

Category: Policies/Procedures/Tools
Q70 - This utility has an effective process in place to deal with managing and supporting change

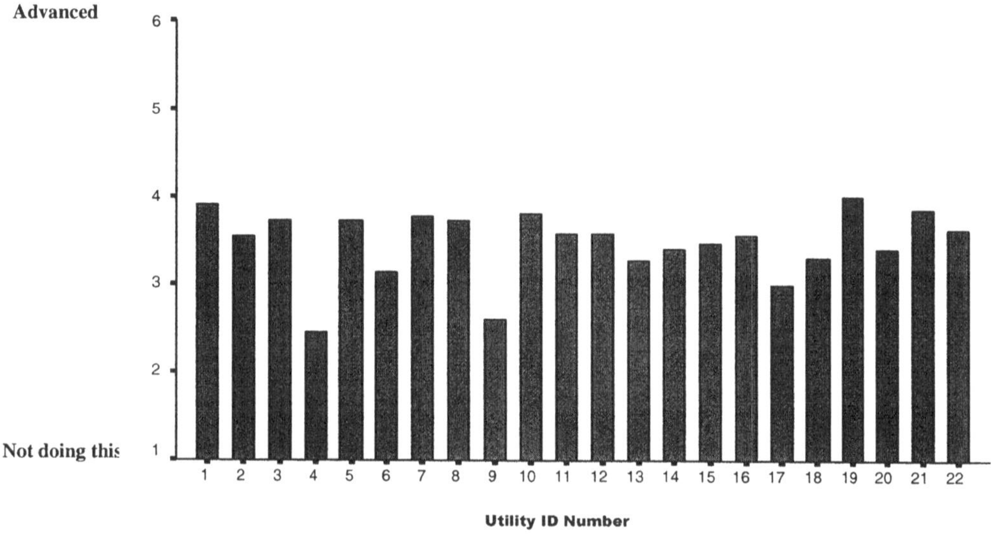

Overall Statistics			
Mean	3.5	Minimum	1
Median	4.0	Maximum	6
Std. Dev.	1.2	Population	22

Category: Policies/Procedures/Tools
Q76 - The cycle time between identifying the need to hire an employee and completing the hiring process is:

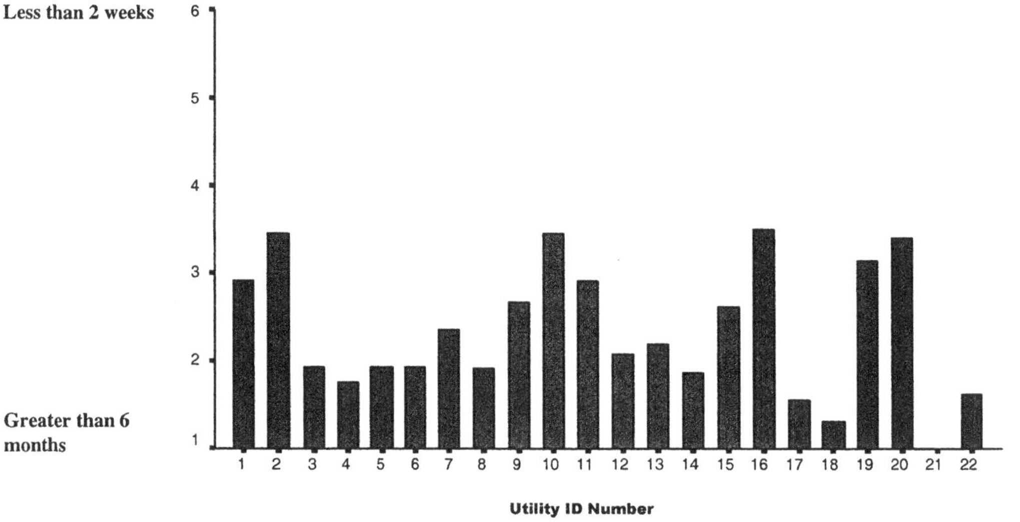

Overall Statistics			
Mean	2.3	Minimum	1
Median	2.0	Maximum	6
Std. Dev.	1.2	Population	22

Category: External Forces
Q38 - External forces create obstacles to effective employment practices

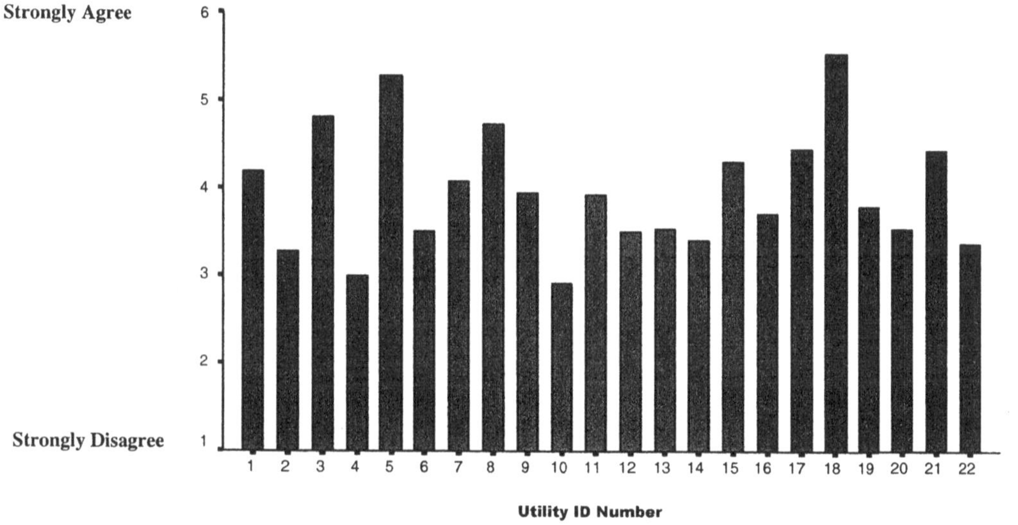

Overall Statistics

Mean	4.0	Minimum	1
Median	4.0	Maximum	6
Std. Dev.	1.3	Population	22

Category: External Forces

Q39 - Introducing today's technology to be more effective is a compelling reason for change at this utility

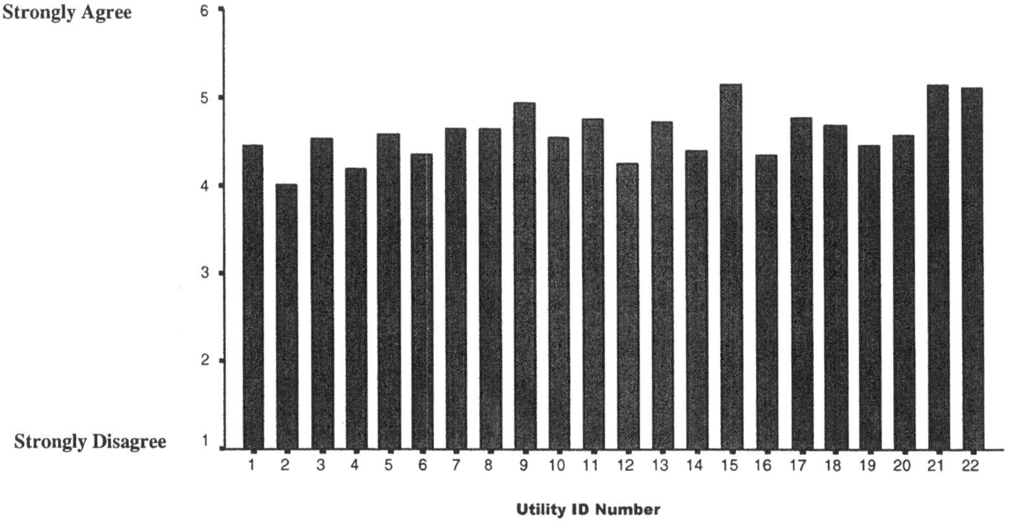

Overall Statistics			
Mean	4.6	Minimum	2
Median	5.0	Maximum	6
Std. Dev.	1.0	Population	22

Category: External Forces

Q40 - Employment practices at this utility include factors other than an employee's skills, abilities, or job performance

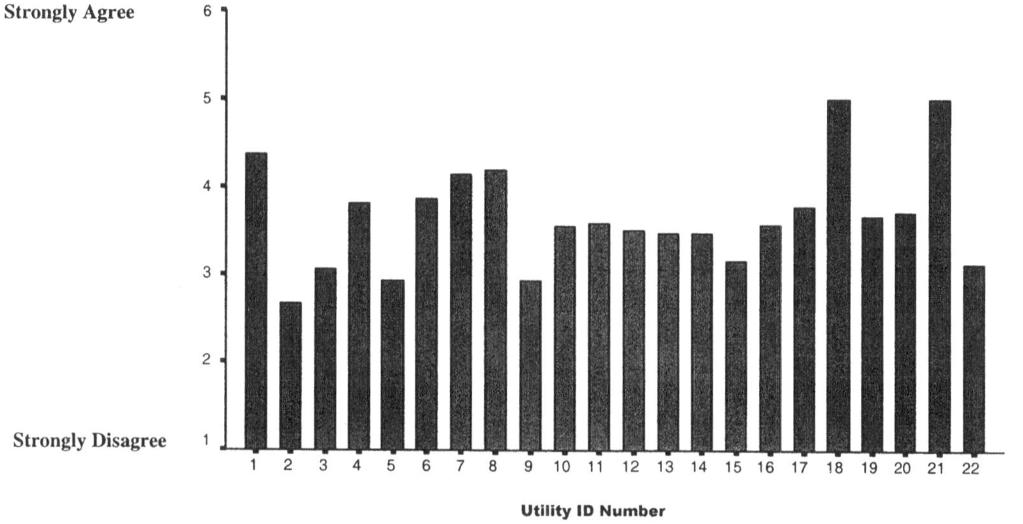

Overall Statistics			
Mean	3.7	Minimum	1
Median	4.0	Maximum	6
Std. Dev.	1.5	Population	22

Category: External Forces

Q45 - This utility places significant emphasis on compliance with governmental regulations regarding personal and public safety, well-being, and workplace environment

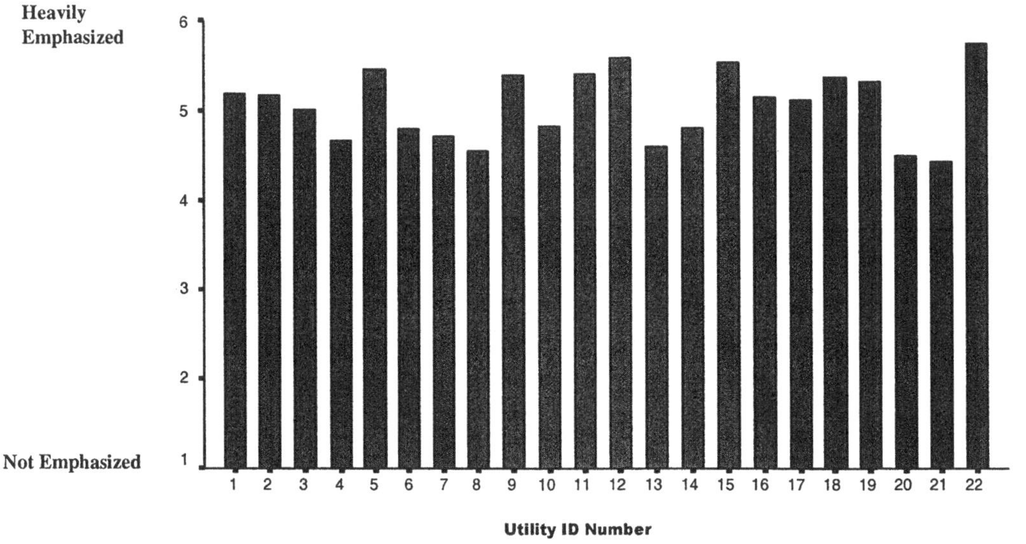

Overall Statistics			
Mean	5.1	Minimum	2
Median	5.0	Maximum	6
Std. Dev.	.9	Population	22

Category: External Forces

Q71 - Effective performance information systems are used at this utility to enable continuous improvement

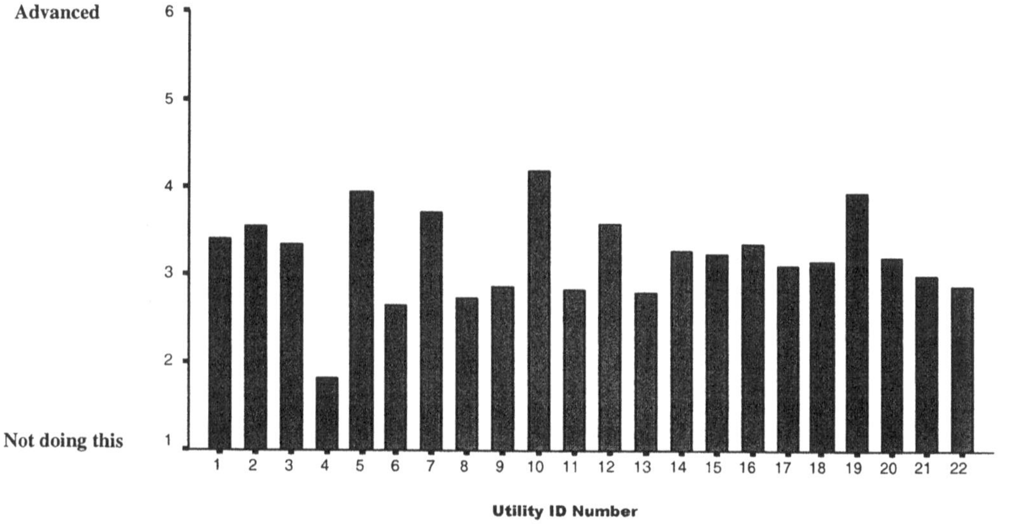

Overall Statistics			
Mean	3.2	Minimum	1
Median	3.0	Maximum	6
Std. Dev.	1.3	Population	22

Category: External Forces

Q86 - The governing body of this utility is an important factor which drives major change in this utility

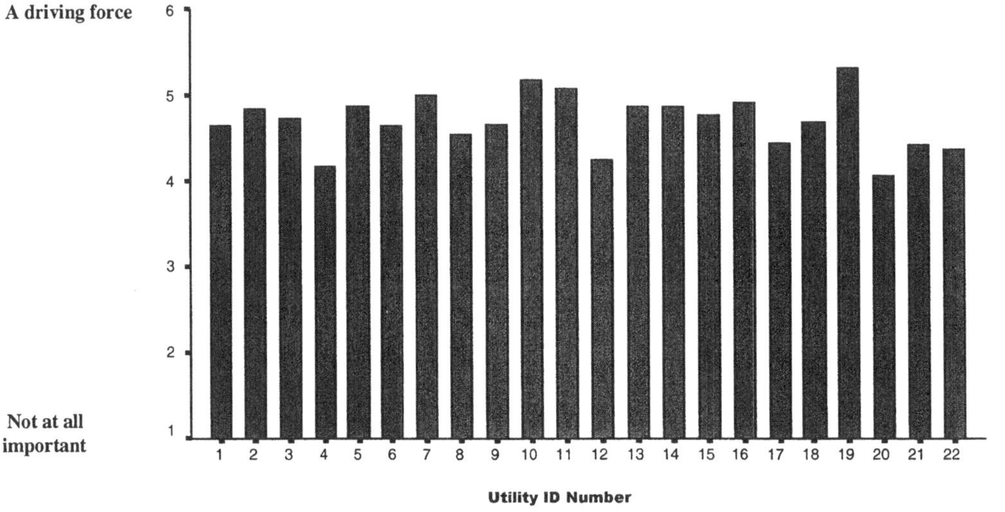

Overall Statistics			
Mean	4.7	Minimum	2
Median	5.0	Maximum	6
Std. Dev.	1.0	Population	22

REFERENCES

American Productivity & Quality Center (APQC) International Benchmarking Clearinghouse. 1999. *Achieving Organizational Excellence Through the Performance Measurement System. Consortium Benchmarking Study.* APQC

American City and County. 1998. Team Re-engineers City's Water Department.

Bennis, Warren. 1997. *Managing People Is Like Herding Cats.* Provo, Utah: Executive Excellence Publishing.

Bennis, Warren. 1989. *On Becoming a Leader.* Reading Mass.: Addison-Wesley Publishing Co.

Bridges, William. 1991. *Managing Transitions: Making the Most of Change.* Reading, Mass.: Addison-Wesley Publishing Co.

Carver, John. 1997. *Boards That Make A Difference: A New Design For Leadership in Nonprofit and Public Organizations.* Jossey-Bass Inc., Publishers

Case, John. 1998. HR Learns How to Open the Books. *HR Magazine.*

Connors, Roger, Tom Smith, and Craig Hickman. 1994. *The Oz Principle: Getting Results Through Individual and Organizational Accountability.* Paramus, NJ: Prentice Hall Press.

Conrad, Charles. *Strategic Organizational Communication.* Orlando, Fla: Harcourt Brace. 1994.

Edwards, Gregg and Snyder, David Pearce. 2000. *Roller Coaster 2000: Forces Re-Shaping Daily Life and Work in America.* The Snyder Family Enterprise.

Hay Group, The: Thomas P. Flannery, David A. Hofrichter, Paul E. Platten. 1996. *People, Performance, and Pay: Dynamic Compensation for Changing Organizations.*

Herman, Roger E. and Gioia, Joyce L. 1998. Making Work Meaningful: Secrets of the Future-Focused Corporation. *The Futurist.*

Jurow, Doree. 1998. Build Management Teams with the Power to Lead. *Power.*

Kanter, Rosabeth Moss, B. A. Stein, R. D. Jick. 1992. *The Challenge of Organizational Change.* New York: The Free Press.

Kaplan, Robert S. and David P. Norton. 1996. *The Balanced Scorecard.* Harvard Business School Press.

Katzenbach, Jon R., and Douglas K. Smith. 1993. *The Wisdom of Teams.* New York: McKinsey & Company, Inc.

Kotter, John P. 1999. *John P. Kotter on What Leaders Really Do.* Boston, Mass.: Harvard Business School Press.

Kotter, John P. 1996. *Leading Change.* Boston, Mass.: Harvard Business School Press.

Labovitz, George, and Victor Rosansky. 1997. *The Power of Alignment: How Great Companies Stay Centered and Accomplish Extraordinary Things.* New York: John Wiley and Sons, Inc.

Lawler, Edward E., III. 1990. *Strategic Pay: Aligning Organizational Strategies and Pay Systems.*

McLagan, Patricia, and Peter Krembs. 1995. *On-The-Level: Performance Communication That Works.*

Osborne, D., and P. Plastrik. March 9 1997. The Union Solution. *The Washington Post,* W08.

Scott-Morgan, Peter. 1994. *The Unwritten Rules of the Game: Master Them, Shatter Them, and Break Through the Barriers to Organizational Change.*

Senge, Peter, Art Kleiner, Charlotte Roberts, Richard Ross, George Roth, and Bryan Smith. 1999. *The Dance of Change*. London, England: Doubleday/Currency, a division of The Doubleday Broadway Publishing Group, a division of Random House, Inc., and Nicholas Brealey Publishing.

Tannenbaum, Robert, and Warren H. Schmidt. 1973. *Harvard Business Review (HBR) Classic.* Number 73311. Boston, Mass.: Harvard College.

U.S. Department of Labor. 1999. *Futurework: Trends and Challenges for Work in the 21st Century.*

U.S. Department of Labor, *Working Together for Public Service*, 1996.

Wheatley, Margaret J. 2000. *Lessons in Leadership Series: Leadership from the Ground Up.*

www.aes.org

www.afscme.org

www.apqc.org

ABBREVIATIONS AND ACRONYMS

ABC	Activity Based Costing
AES	AES Corporation
AFSCME	American Federation of State, County and Municipal Employees
APQC	American Productivity and Quality Center
ATF	Apprenticeship Task Force
AWWA	American Water Works Association
AwwaRF	Awwa Research Foundation
BSC	Balanced Scorecard
CIS	Customer Information System
CN	Canada
CPA	Certified Public Accountant
DPT	Decision Partner Team
GAAP	Government Accounting and Auditing Principles
GARM	Goals, Actions, Results, and Measures
GIS	Geographical Information System
HBR	Harvard Business Review
HR	Human Resources
HRMS	Human Resources Management System
ISO	International Standards Organization
KRA	Key Results Areas
KY	Kentucky
LWC	Louisville Water Company
NWT	Natural Work Teams
OCSEA	Ohio Civil Service Employees Association
OR	Oregon
PAC	Project Advisory Committee
PALM	Participative Association of Labor and Management
PWSD	Phoenix Water Services Department
SEIU	Sanitation Employees International Union
SPU	Seattle Public Utilities
US	United States

WA	Washington
WPPM	Work Planning and Performance Management